Lecture-Tutorials for Introductory Astronomy

PRELIMINARY EDITION

Jeffrey P. Adams
Montana State University

Edward E. Prather
University of Arizona

Timothy F. Slater
University of Arizona

and the

Conceptual Astronomy and Physics Education Research (CAPER) Team

Prentice Hall
Pearson Education, Inc.
Upper Saddle River, New Jersey 07458

Senior Editor: *Erik Fahlgren*
Assistant Editor: *Christian Botting*
Production Editor: *Donna Young*
Assistant Managing Editor, Science: *Beth Sweeten*
Manufacturing Manager: *Trudy Pisciotti*
Manufacturing Buyer: *Ilene Kahn*
Cover Design: *Bruce Kenselaar*

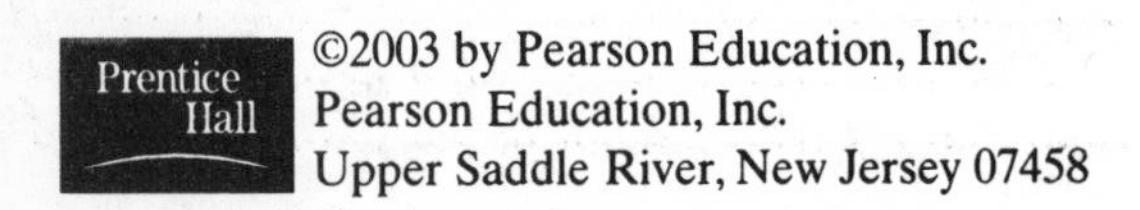

Pearson Education, Inc.
Upper Saddle River, New Jersey 07458

Printed in the United States of America
10 9 8 7 6 5 4

ISBN 0-13-101109-X

Pearson Education LTD., *London*
Pearson Education Australia PTY. Limited, *Sydney*
Pearson Education Singapore, Pte. Ltd.
Pearson Education North Asia Ltd, *Hong Kong*
Pearson Education Canada, Ltd., *Toronto*
Pearson Educacion de Mexico, S.A. de C.V.
Pearson Education—Japan, *Tokyo*
Pearson Education Malaysia, Pte. Ltd.

TABLE OF CONTENTS

Instructor's Preface

Each year, over 200,000 students take introductory astronomy—hereafter referred to as ASTRO 101; the majority of these students are non-science majors. Most are taking ASTRO 101 to fulfill a university science requirement and many approach science with some mix of fear and disinterest. The traditional approach to winning over these students has been to emphasize creative and engaging lectures, taking full advantage of both demonstrations and awe-inspiring astronomical images. However, what a growing body of evidence in astronomy and physics education research has been demonstrating is that even the most popular and engaging lectures do not engender the depth of learning for which faculty appropriately aim. Rigorous research into student learning tells us that one critical factor in promoting classroom learning is students' active "minds-on" participation. This is best expressed in the mantra: "It's not what the teacher does that matters; it's what the students do."

Lecture-Tutorials for Introductory Astronomy has been developed in response to the demand from astronomy instructors for easily implemented student activities for integration into existing course structures. Rather than asking faculty—and students—to convert to an entirely new course structure, our approach in developing *Lecture-Tutorials* was to create classroom-ready materials to augment more traditional lectures. Any of the activities in this manual can be inserted at the end of lecture presentations and, because of the education research program that led to the activities' development, we are confident in asserting that the activities will lead to deeper and more complete student understanding of the concepts addressed.[1]

Each *Lecture-Tutorial* presents a structured series of questions designed to confront and resolve student difficulties with a particular topic. Confronting difficulties often means answering questions incorrectly; this is expected. When this happens, the activities are crafted to help a student understand where her or his reasoning went wrong and to develop a more thorough understanding as a result. Therefore, while completing the activities, students are encouraged to focus more on their reasoning and less on trying to guess an expected answer. The activities are meant to be completed by students working in pairs who "talk out" the answers with each other to make their thinking explicit.

At the conclusion of each *Lecture-Tutorial*, instructors are strongly encouraged to engage their class in a brief discussion about the particularly difficult concepts in the activity—an essential implementation step that brings closure to the activity. The online *Instructor's Guide* also provides "post-tutorial" questions that can be used to gauge the effectiveness of the *Lecture-Tutorial* before moving on to new material.

Acknowledgments

Lecture-Tutorials for Introductory Astronomy was developed by the Conceptual Astronomy and Physics Education Research (CAPER) Team at Montana State University and the University of Arizona with generous support from the National Science Foundation (NSF CCLI #9952232 and NSF Geosciences Education #9907755). Gina Brissenden provided initial project definition, management, and graphical layout efforts. Jack Dostal and Larry Watson assisted directly in writing activities. Numerous other individuals contributed to this project through critical assessment and the national field-testing of the materials. These individuals include Chija Skala Bauer, Tom Brown, Dave Bruning, Beth Hufnagel,

---

[1] Instructors can go to http://www.prenhall.com/tiponline for online *Instructor's Guide* that gives detailed information on classroom implementation as well as evidence of the efficacy of specific activities.

Lauren Jones, Janet Landato, Ed Murphy, and Erika Offerdahl. Particularly noteworthy were the extensive reviews by Steve Shawl and Janelle Bailey that continually kept us on our toes. In addition we must thank Christian Botting who helped us with day-to-day publication issues and Alison Reeves who repeatedly encouraged us to continue and helped us frame the initial ideas for this work. Most importantly, we wish to express our appreciation to our students who patiently endured early versions of these tutorials and unselfishly provided extensive feedback.

Note to the Student

Welcome to the study of astronomy! You are about to embark on a grand study of the cosmos. To help you better understand the topics of your course we have created this series of activities called *Lecture-Tutorials*. In each activity, you are asked a short series of questions that will require you to work in collaboration with your classmates to help you learn important and difficult concepts in astronomy. For every question in these activities it is important that you write out a detailed answer. This is critical because you will certainly be using these materials to study for exams. It is also important because part of the learning process is being able to express complex ideas in writing.

We strongly encourage you to actively engage in completing these activities in collaboration with another student. The process of deciphering the questions and negotiating a common language to write your answers will help you understand the concepts more deeply. Specifically, the *Lecture-Tutorials* are designed to give you a starting point to think carefully and talk with others about concepts in astronomy. Above all, have fun exploring astronomy!

J.P. Adams, E.E. Prather, T.F. Slater and the
Conceptual Astronomy and Physics
Education Research (C.A.P.E.R.) Team

Two stars, A and B, are each shown at four different times (1, 2, 3, and 4) as the celestial sphere rotates about the stationary Earth. Figure 1 shows the entire celestial sphere and your location in the northern hemisphere. Figure 2 shows only the portion of the celestial sphere above your horizon, which is all you can see.

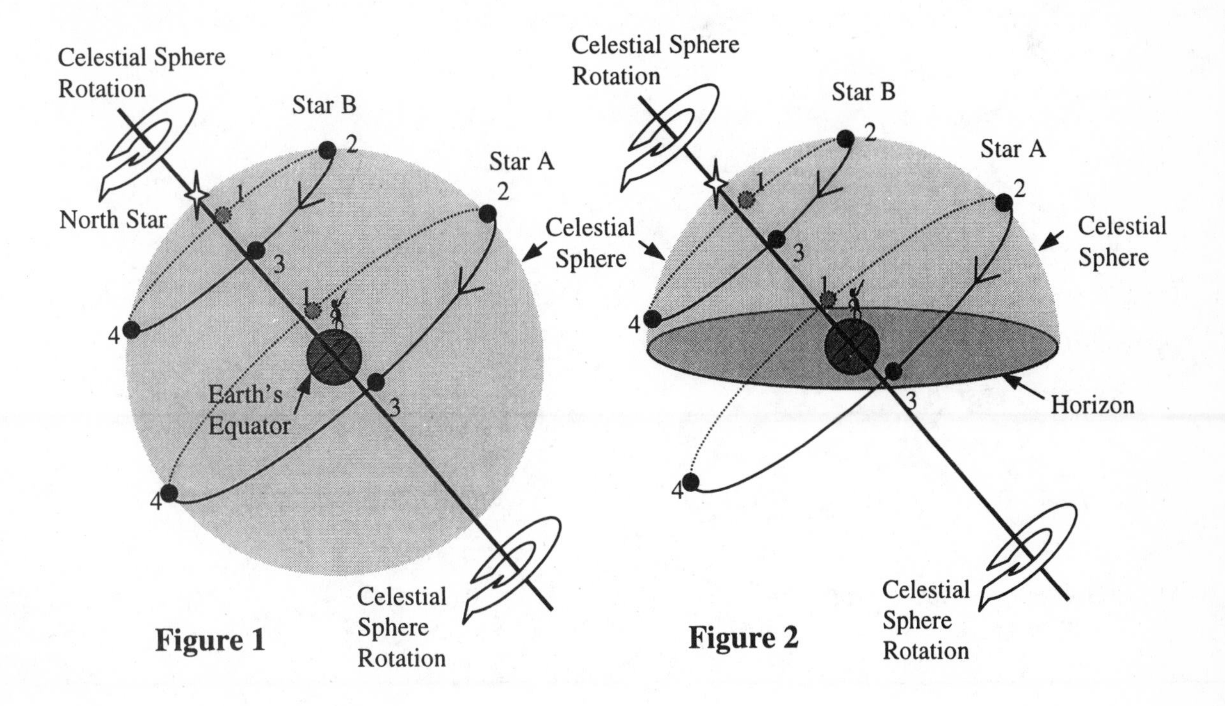

1) Star A is just visible above your eastern horizon at position 1. At which of the numbered positions is it just visible above your western horizon?

2) At what position(s), if any, does Star B rise and set?

3) Two students are discussing their answers to question 2.

Student 1: *Locations B1 and B3 are on my horizon because they are rising and setting just like A1 and A3.*

Student 2: *Figure 2 shows that Star B is lowest in the sky at B4 and is just above the northern horizon. Star B never goes below the horizon.*

Do you agree with Student 1, Student 2, both, or neither? Why?

4) For each indicated position, describe where in the sky you must look to see the star at that time. Each description requires two pieces of information: the direction you must face (north, northeast, east, etc.) and how far above the horizon you must look (low, high, or directly overhead). If you cannot see the star, state that explicitly. The descriptions for four positions are given as examples.

   a) A1: *east, low*

   b) A2:

   c) A3:

   d) A4:

   e) North Star: *north, high*

   f) B1:

   g) B2: *directly overhead (only one star can be directly overhead)*

   h) B3: *northwest, high*

   i) B4:

   Check your answers with a nearby group and resolve any inconsistencies.

5) Does Star B ever set? Explain.

Part I: Looking North

You are in the northern hemisphere at 6 PM. Looking north, the sky appears as shown in Figure 1. The positions and motions of the stars in Figure 1 can be understood by imagining yourself as the observer at the center of the **celestial sphere** as shown in Figure 2. In the celestial sphere model, the Earth is stationary and the stars are carried on a sphere that rotates about an axis through the North Star.

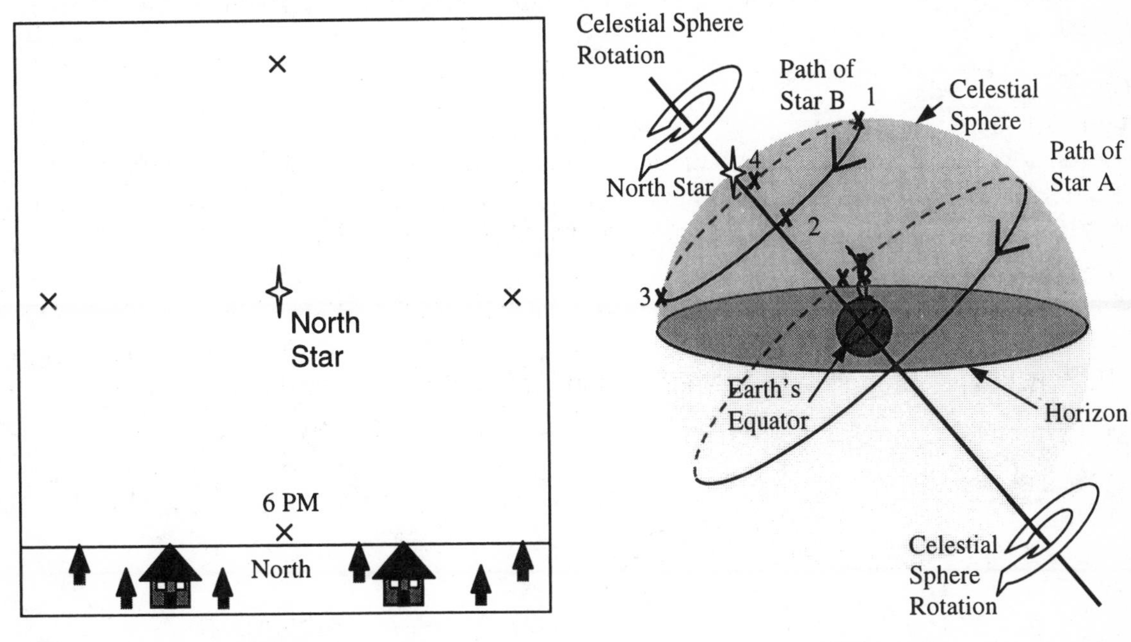

Figure 1 **Figure 2**

The x's in both figures represent four of the positions through which star B will pass during the course of one revolution of the celestial sphere. Ignore star A until question 5.

1) Note in Figure 1 the position of Star B at 6 PM. Circle the numbered position (1, 2, 3, or 4) in Figure 2 that corresponds to the location of star B at 6 PM.

2) The rotation of the celestial sphere carries star B around so that it returns to the same position at about 6 PM the next evening. Label each of the x's in both figures with the approximate time at which star B will arrive (e.g., the location you circled in question 1 will be labeled "6 PM").

3) Using Figure 2, describe the direction you have to look to see Star B at 6 AM.

4) The position directly overhead is called **zenith**. Label the direction of the zenith on Figure 1. How does the direction of the zenith compare to the direction that you identified in question 3?

5) Using Figure 2, describe in words the position of Star A half way between rising and setting.

6) In Figure 1, draw a straight arrow to represent the direction Star B would be moving when at each of the four locations marked with an x. Check your answers with a nearby group.

Part II: Looking East

Figure 3 extends your view along the eastern horizon showing the positions of stars A and B at 6 PM. The arrow shown is provided to indicate the direction that Star B will be moving at 6 PM.

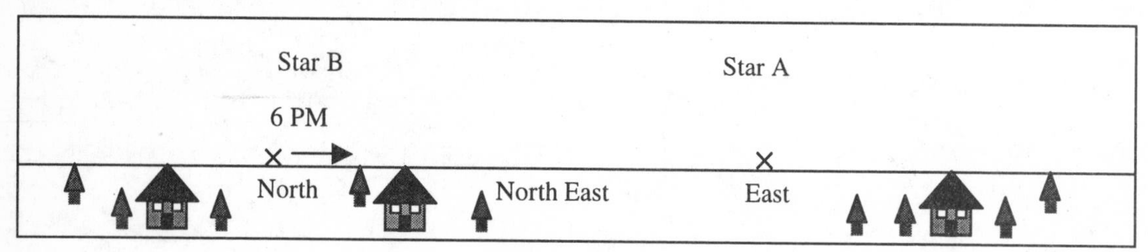

Figure 3

7) Recall that in question 5, you found that Star A ends up high in the southern sky half way between rising and setting (and therefore never passes through your zenith). Draw a straight arrow at the x in the east in Figure 3 (the position of Star A at 6 PM) to indicate the direction Star A moves as it rises. Studying Figure 2 can also help clarify your answer.

8) Two students are discussing the direction of motion of a star rising directly in the east.

Student 1: *Stars move east to west so any star rising directly in the east must be moving straight up so that it can end up in the west. If the arrow were angled, the star would not set in the west.*

Student 2: *I disagree. From Figure 2, the path of Star A starts in the east, swings through the southern sky yet still sets in the west. To do this it has to move toward the south as it rises so I drew my arrow angled to the right.*

Do you agree with Student 1, Student 2, both, or neither? Why?

9) Consider the question and student response below.

Question: Your are standing outside looking toward the north and see a star directly below the north star and just above your horizon. After 15 minutes, in what direction will the star have moved?

Student: *The star will have moved to the west, which is my left. All stars move from east to west.*

Figures 1 and 3 show that a star at this location would be moving to the east. What doesn't the student understand about the statement, "stars move east to west"?

Part I: Monthly Differences

Figure 1 shows a heliocentric, perspective view of the Earth-Sun system indicating the direction of both the daily rotation of the Earth about its own axis and its annual orbit about the Sun. You are the observer shown in Figure 1, located in the northern hemisphere. You are facing toward the southern horizon.

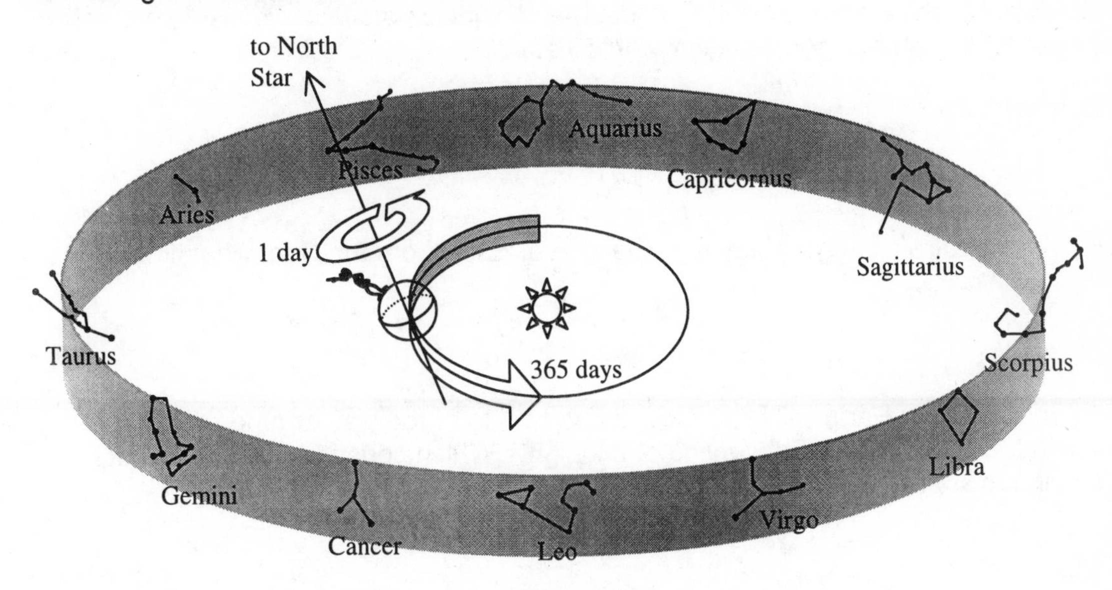

Figure 1

Figure 2 shows a horizon view of what you would see when facing south on this night at the same time as shown in Figure 1.

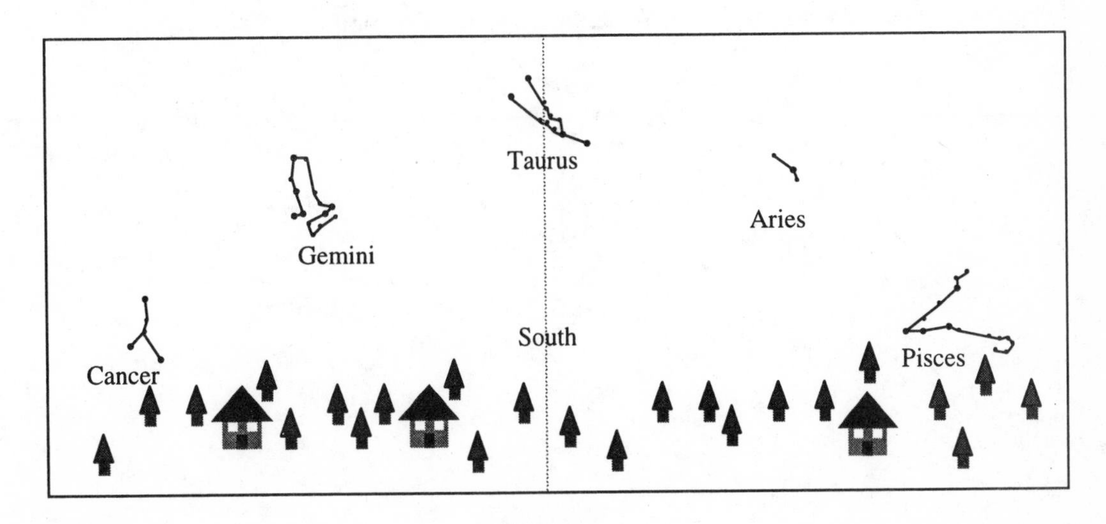

Figure 2

1) Which labeled constellation do you see highest in the southern sky?

2) What constellation is just to the left (i.e., east) and what constellation is just to the right (i.e., west) of the highest constellation at this instant?
left: right:

3) Noting that you are exactly on the opposite side of Earth from the Sun, what time is it?

4) One month later the Earth will have moved one-twelfth of the way around the Sun. You are again facing south while observing at midnight. Which constellation will now be highest in the southern sky?

5) Do you have to look east or west of the highest constellation that you see now to see the constellation that was highest one month ago?

6) Does the constellation that was highest in the sky at midnight a month ago now rise earlier or later than it rose last month? Explain your reasoning.

Part II: Daily Differences

Figure 3 shows the same Earth-Sun view including the bright star Betelgeuse, which is between Taurus and Gemini.

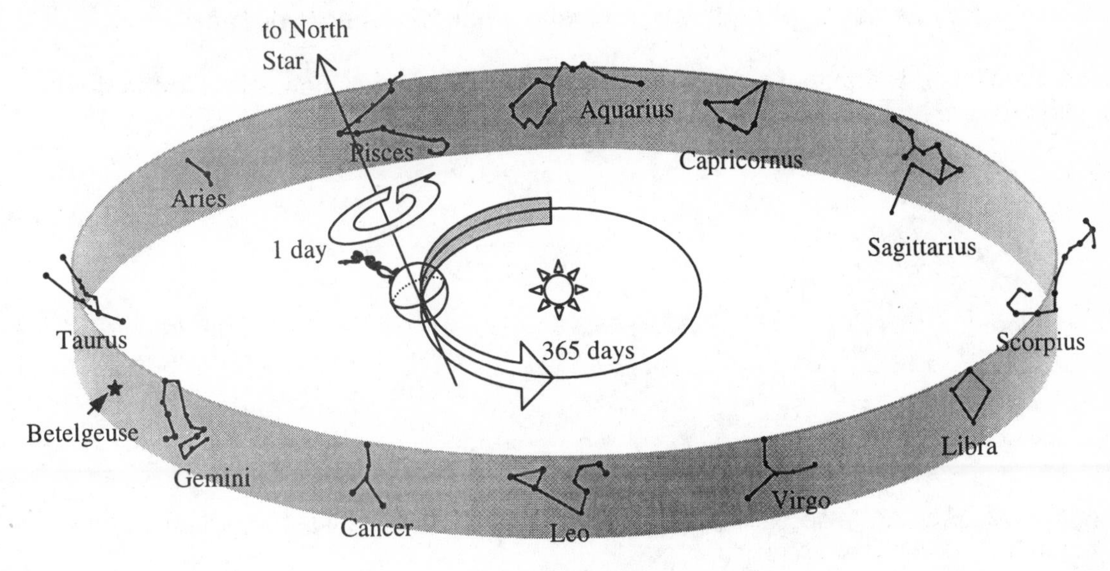

Figure 3

7) Last night you saw the star Betelgeuse exactly on your eastern horizon at 5:47 PM. At 5:47 tonight, will Betelgeuse be above, below, or exactly on your eastern horizon?

8) Two students are discussing their answers to question 7.

Student 1: *The Earth makes one complete rotation about its axis each day so Betelgeuse will rise at the same time every night. It will therefore be exactly on the eastern horizon.*

Student 2: *No. The constellation Taurus rises earlier each month and so it must rise a little bit earlier each night. Betelgeuse must do the same thing. Tonight it would rise a little before 5:47 and be above the eastern by 5:47. You are confusing the sidereal and solar day.*

Do you agree with Student 1 or Student 2? Why?

9) How long did it take for Betelgeuse to return to exactly the same position it was on the previous night—slightly less than 24 hours, exactly 24 hours, or slightly more than 24 hours?

10) The celestial sphere in Figure 4 helps us picture the motion of the night sky from a stationary Earth by imagining the sphere to rotate approximately once every day. To make this model consistent with your answer to question 9, how much time should the celestial sphere take to complete one revolution about a fixed Earth—slightly less than 24 hours, exactly 24 hours, or slightly more than 24 hours? Check your answer with a nearby group.

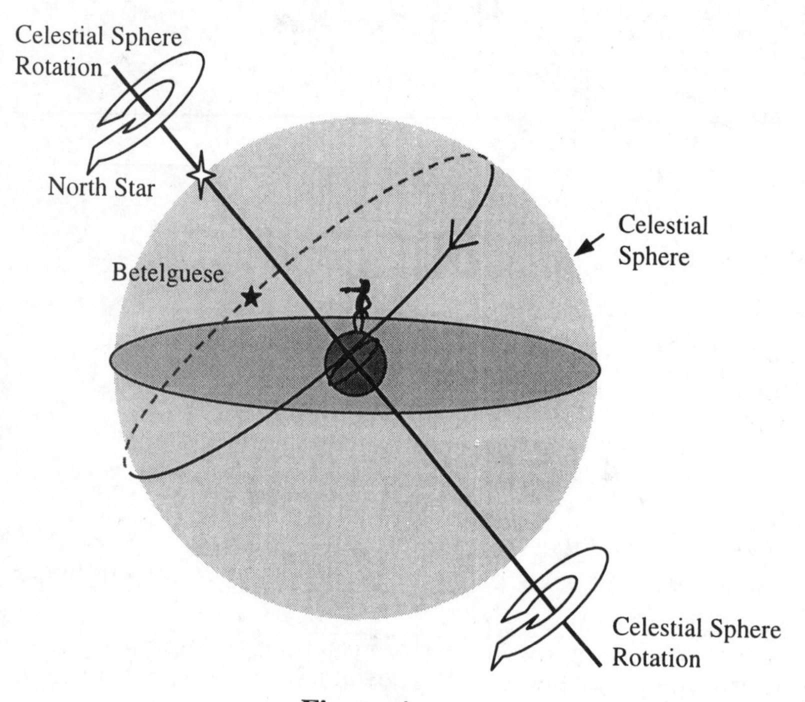

Figure 4

Part I: Solar Day

Figure 1 shows a top-down view of the Earth-Sun system. Arrows indicate the directions of the rotational and orbital motion of Earth. For the observer shown, the Sun is highest in the sky at 12 noon.

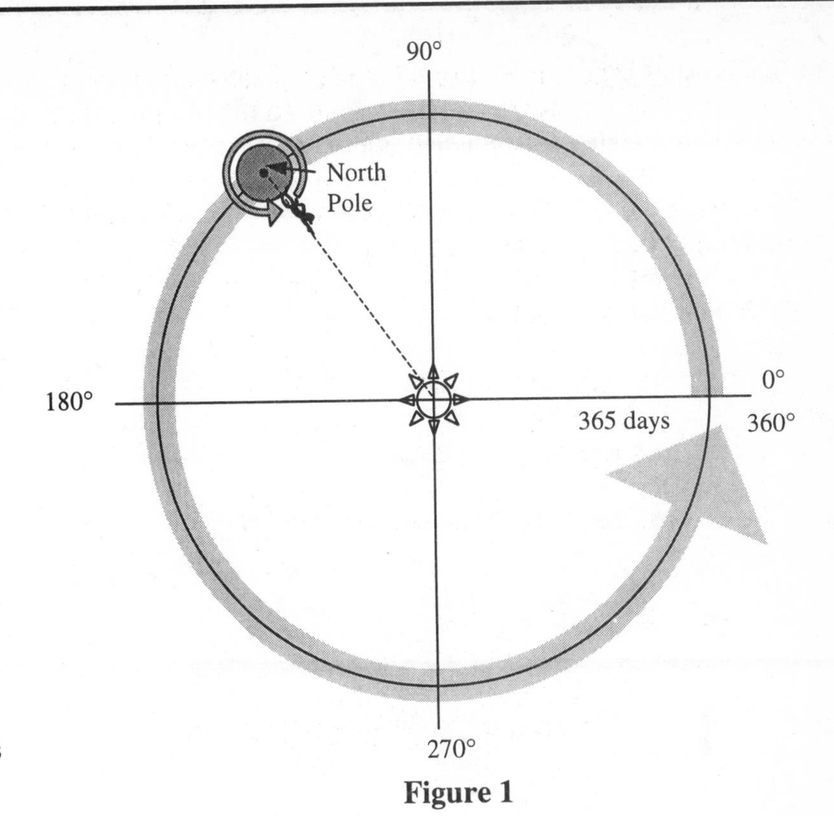

Figure 1

1) Earth orbits the Sun in a counterclockwise direction once every 365 days. Approximately how many degrees does Earth move along its orbit in one day?

2) As Earth orbits the Sun, it also rotates in a counterclockwise direction about its axis as shown in Figure 1. We define 24 hours as the time from when the Sun is highest in the sky one day to when it is highest in the sky the next day. How many degrees does Earth rotate about its axis in exactly 24 hours: 360°; slightly less than 360°; or slightly more than 360°?

3) Two students are discussing their answer to question 2.

Student 1: *Earth rotates about its axis once every 24 hours and one rotation equals 360°.*

Student 2: *No. When Earth has gone around 360° it has also moved a small amount counterclockwise around the Sun, which means the Sun is not yet at its highest point. Earth must spin a little bit more for the Sun to reach its highest point.*

With which student do you agree? Explain your answer.

Part II: Sidereal Day

We define a **solar day** as the time it takes for the Sun to go from highest in the sky to highest in the sky and we divide that time into 24 hours.

A **sidereal day** is defined as the time it takes for Earth to rotate *exactly* 360° about its axis with respect to the stars. Since Earth rotates more than 360° in a solar day, a sidereal day is about 4 minutes shorter than a solar day.

Imagine that some time in the future the direction that Earth orbits the Sun is somehow reversed. Earth now orbits approximately 1° *clockwise* each day. However, the rotation about its own axis remains counterclockwise at the same rate.

4) In the space below, create a sketch similar to Figure 1 to depict this situation.

5) Through how many degrees will Earth now rotate in a sidereal day?

6) Through how many degrees will Earth now rotate in a solar day?

7) Which is longer, the solar or the sidereal day?

8) Is a sidereal day now longer, shorter, or the same length as a sidereal day before Earth's orbit changed?

9) Is a solar day now longer, shorter, or the same length as a solar day before Earth's orbit changed?

For all parts of this activity, it is helpful to imagine that the stars are so bright (or our Sun so dim) that the stars can be seen during the day so that your sky appears as in Figure 1.

Part I: Daily Motion

On December 1, at noon, you are looking toward the south and see the Sun among the stars of the constellation Scorpius as shown in Figure 1.

1) At 3 PM this afternoon, will the Sun appear among the stars of the constellation Sagittarius, Scorpius, or Libra?

2) Two students are discussing their answers to question 1.

   Student 1: *The Sun moves from east to west across the sky. By 3 PM it will have moved to the west into the constellation Libra.*

   Student 2: *You're forgetting that the stars and constellations also move to the west like the Sun. So, the Sun will still be in Scorpius at 3 PM.*

   Do you agree with Student 1 or Student 2? Why?

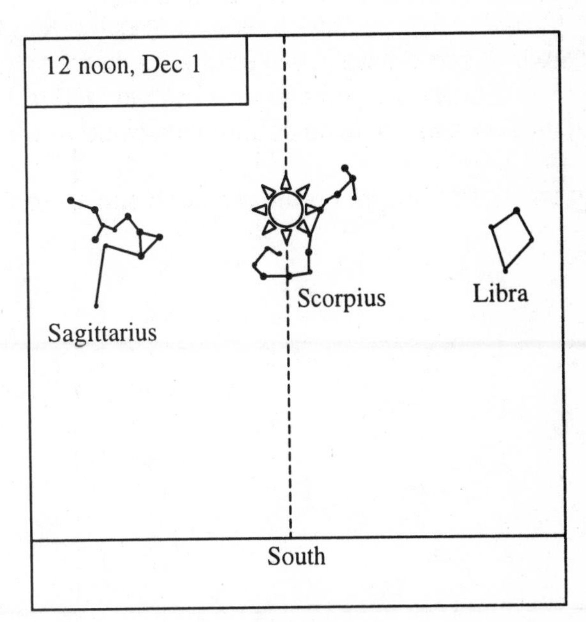

Figure 1

In the **celestial sphere model**, the stars' daily motions result from the rotation of the celestial sphere.

3) Can we understand the Sun's **daily motion** by assuming that the Sun is at a fixed position on the celestial sphere (in this case in Scorpius) and is carried along its path by the sphere's rotation?

Part II: Monthly Changes

By careful observation we find that the celestial sphere rotates slightly more than 360° every 24 hours. Figure 2 shows the same view of the sky (as Figure 1) on **December 2 at noon.** For comparison, the view from the previous day at the same time is also shown in gray.

4) Draw the location of the Sun as accurately as possible in Figure 2.

5) Figure 3 shows the same view of the sky (as Figure 1) one month later on **January 1 at noon**. Draw the location of the Sun as accurately as possible in this figure.

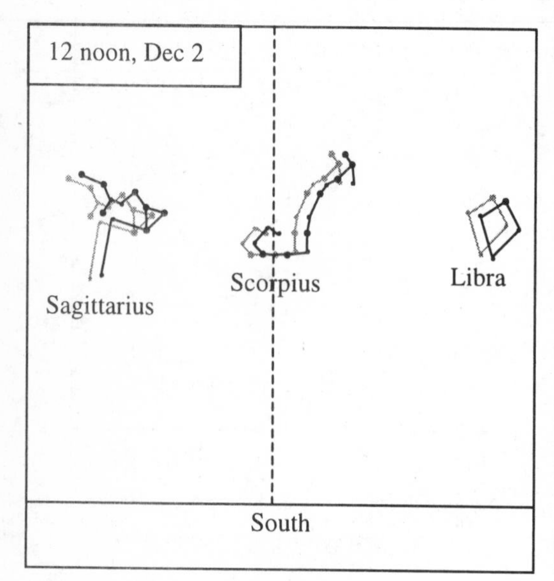

Figure 2

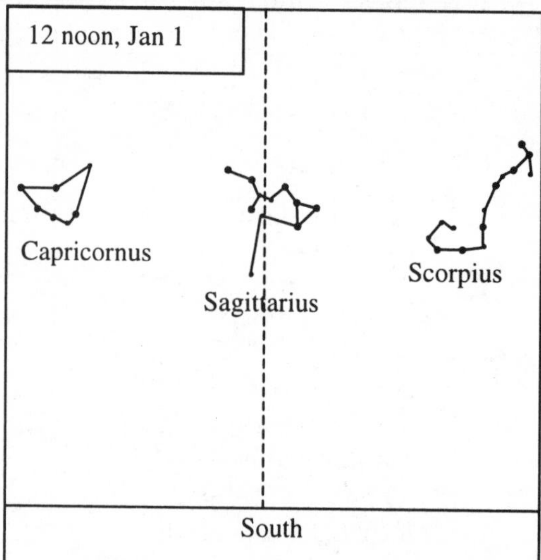

Figure 3

6) Two students are discussing their answers to questions 4 and 5.

Student 1: *The Sun will always lie along the dotted line in the figures when it's noon.*
Student 2: *But, we saw in question 3 that the Sun's motion can be modeled by assuming it is stuck to the celestial sphere. The Sun must, therefore, stay in Scorpius.*
Student 1: *If that were true then by March the Sun would be setting at noon. The Sun must shift a little along the celestial sphere each day so that in 30 days it has moved into the next constellation.*

Do you agree with Student 1 or Student 2? Why?

7) Why is it OK to pretend that the Sun is fixed on the celestial sphere through the course of a single day as suggested in question 3 even though we know from questions 5 and 6 that the Sun's position is not truly fixed on the celestial sphere?

Part III: The Ecliptic

The zodiacal constellations were of special interest to ancient astronomers because these are the constellations through which the Sun moves throughout the year. This was incorporated in their celestial sphere model by having the Sun loosely fixed to the celestial sphere but allowed to slip a small amount each day. The Sun's position on the celestial sphere on December 1 is represented in Figure 4.

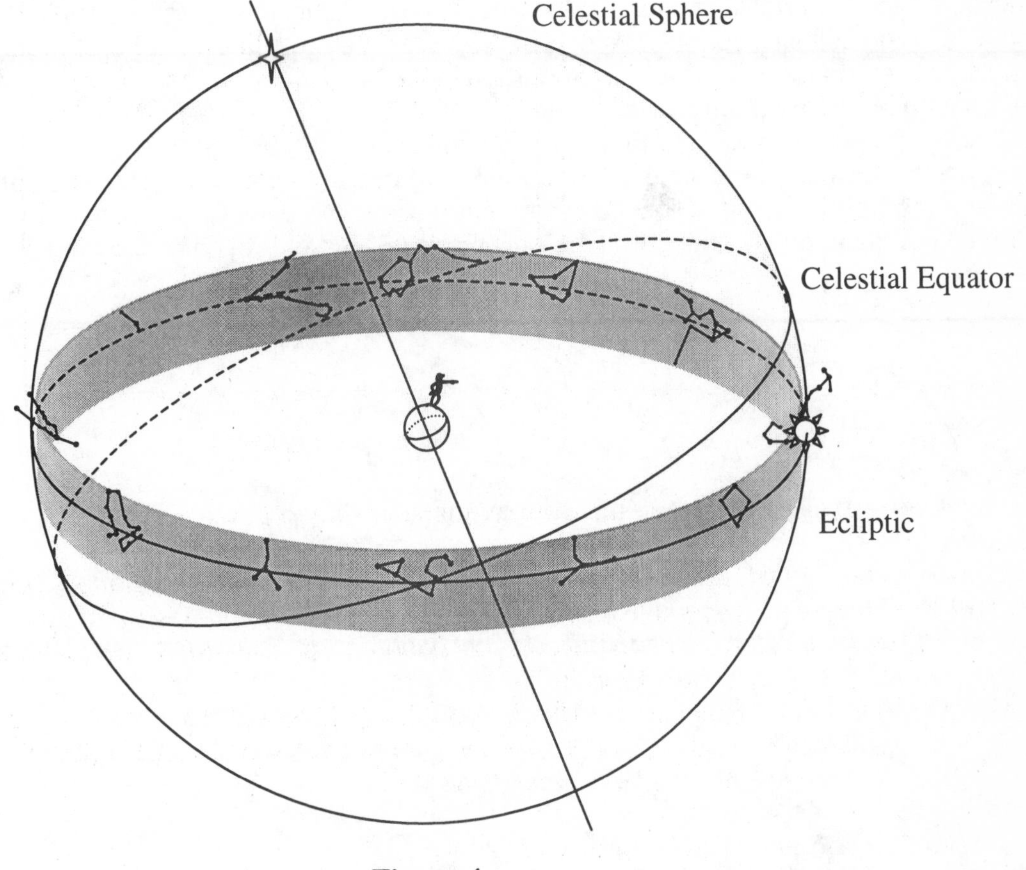

Figure 4

8) Draw the Sun's position on the celestial sphere January 1.

9) Label all of the constellations according to the approximate month that the Sun would be there.

The line drawn through these constellations, tracing out the Sun's annual path, is called the **ecliptic**.

10) About how many times does the celestial sphere rotate in the time it takes the Sun to move between two adjacent constellations (i.e., 1/12 of the way around) along the ecliptic?

11) How long does it take the Sun to make one complete trip around the ecliptic (i.e., from Scorpius to Scorpius)?

Part IV: Wrap Up

It is important to realize that the ecliptic represents an ANNUAL drift of the Sun and does not represent the daily path of the Sun. Instead, the rotation of the celestial sphere is responsible for the Sun's daily motion. Also, since the ecliptic is tilted with respect to the rotation axis of the celestial sphere, the ecliptic "wobbles" as the celestial sphere rotates. The Sun's position on the ecliptic is only important in deciding whether the Sun's daily path will carry it high in the sky (summer) or low in the sky (winter). In Figure 5a the Sun's position along the ecliptic and its path for one day are shown for two different dates: December 1 (in Scorpius), and June 1 (in Taurus). Figures 5b, 5c and 5d show the path of the Sun and the wobble of the ecliptic at 6-hour intervals as the celestial sphere rotates. Study these figures carefully noting that the ecliptic and Sun are both carried by the celestial sphere.

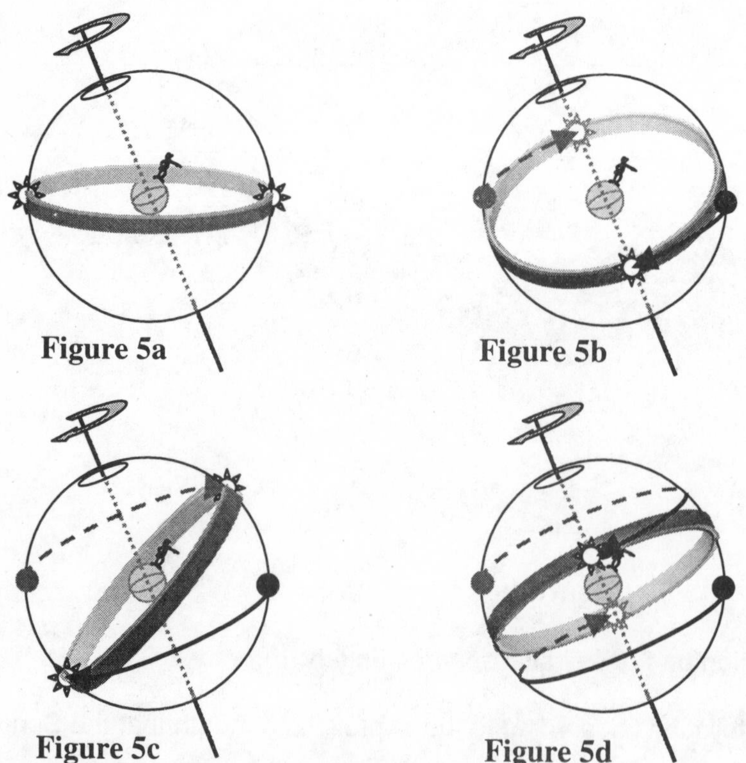

Figure 5a Figure 5b

Figure 5c Figure 5d

12) On Figure 5d, label the ecliptic (Sun's annual path) and the Sun's daily path for December 1 and June 1.

PRELIMINARY EDITION, 2002

Figure1 illustrates the sky as seen from the continental US. It shows that the Sun's daily path across the sky is longest on June 21 and shortest on December 21. In addition, on June 21, which is called the summer solstice, the Sun reaches its highest altitude in the sky above the horizon at about noon. The figure shows that the Sun never actually reaches the zenith for any observer in the continental US. In other words, the Sun is never directly overhead. Over the six months following the summer solstice, the noontime Sun achieves progressively lower and lower altitudes until December 21, the winter solstice. Thus, we see that the path of the Sun through the southern sky changes considerably over the course of a year.

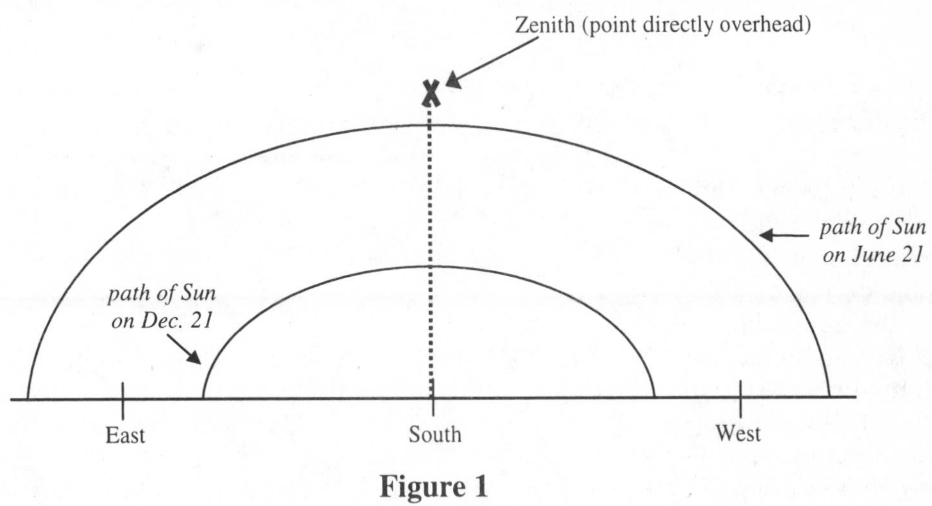

Figure 1

1) According to Figure 1, in which direction would you look to see the Sun at noon today?

   Circle one: East Southeast directly South Southwest West

2) If it is wintertime right now (after the winter solstice), how does the altitude of the Sun at noon change as summer approaches?

   Circle one: increases stays the same decreases

3) If Figure 1 is a reasonable representation for observers in the continental United States, is there ever a time of year when the Sun is directly above us at the zenith (looking straight up) at noon? If so, on what date does this occur?

4) If the Sun rises south of East in December and rises north of East in June, estimate the date(s) when the Sun will rise directly in the East.

5) Does the Sun always set in precisely the same location throughout the year? If not, describe how the location of sunset changes throughout the year.

Figure 2 shows a small, vertical stick, which casts a shadow, sitting on a large piece of paper or poster-board. You can think of this to be somewhat like a sundial.

On two different days of the year, the very top of the shadow is marked with an "*x*" every couple of hours throughout the day. Although this sketch is somewhat exaggerated, these *shadow plots* indicate how the position of the Sun changes in the sky through the course of these two days. The following questions are designed to show the relationship between Figure 1 on the previous page and Figure 2 at right.

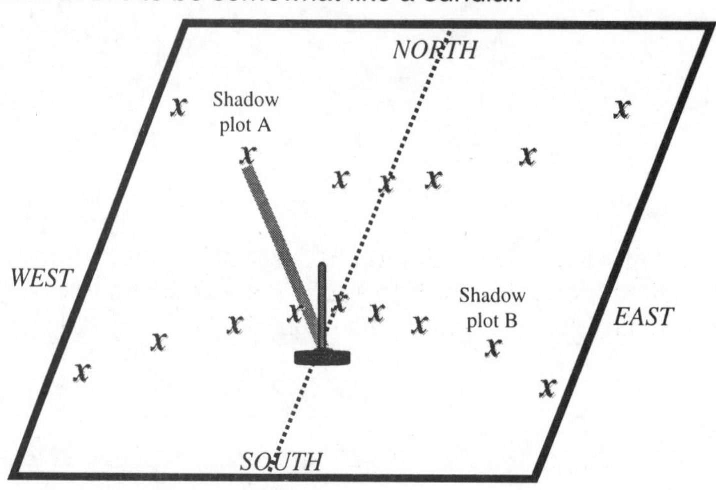

Figure 2

6) Using Figures 1 and 2, if the Sun rises in the Southeast, where would the shadow of the Sun be cast on the poster-board?

Circle one: West Northwest North Northeast East Southeast

7) Clearly circle the "x" for the shadow that corresponds to the time of noon for path A and for path B.

8) Compare the position of the "x" that corresponds to noon for shadow plots A and B. Which shadow plot (A or B) corresponds to a path of the Sun in which the Sun is highest in the sky? Explain your reasoning.

9) Which shadow plot (A or B) most closely corresponds to the Sun's path through the sky during the summer and which corresponds with the winter? Explain your reasoning.

10) Based on the shadow plots in Figure 2, during which time of the year (summer or winter) does the Sun rise south of East? Explain your reasoning.

11) If you were to mark the top of the stick's shadow with an "x," where would the "x" be placed along the north-to-south line to indicate the Sun's position at noon *today*? Clearly explain why you would place the "x" where you did.

12) Is there ever a clear day (no clouds) of the year where the stick casts no shadow? If so, when does this occur and where exactly in the sky does the Sun have to be?

Consider the star map for July at midnight shown in Figure 1. In particular, notice that the directions of north and east have been identified and that the names of different star groups have been provided.

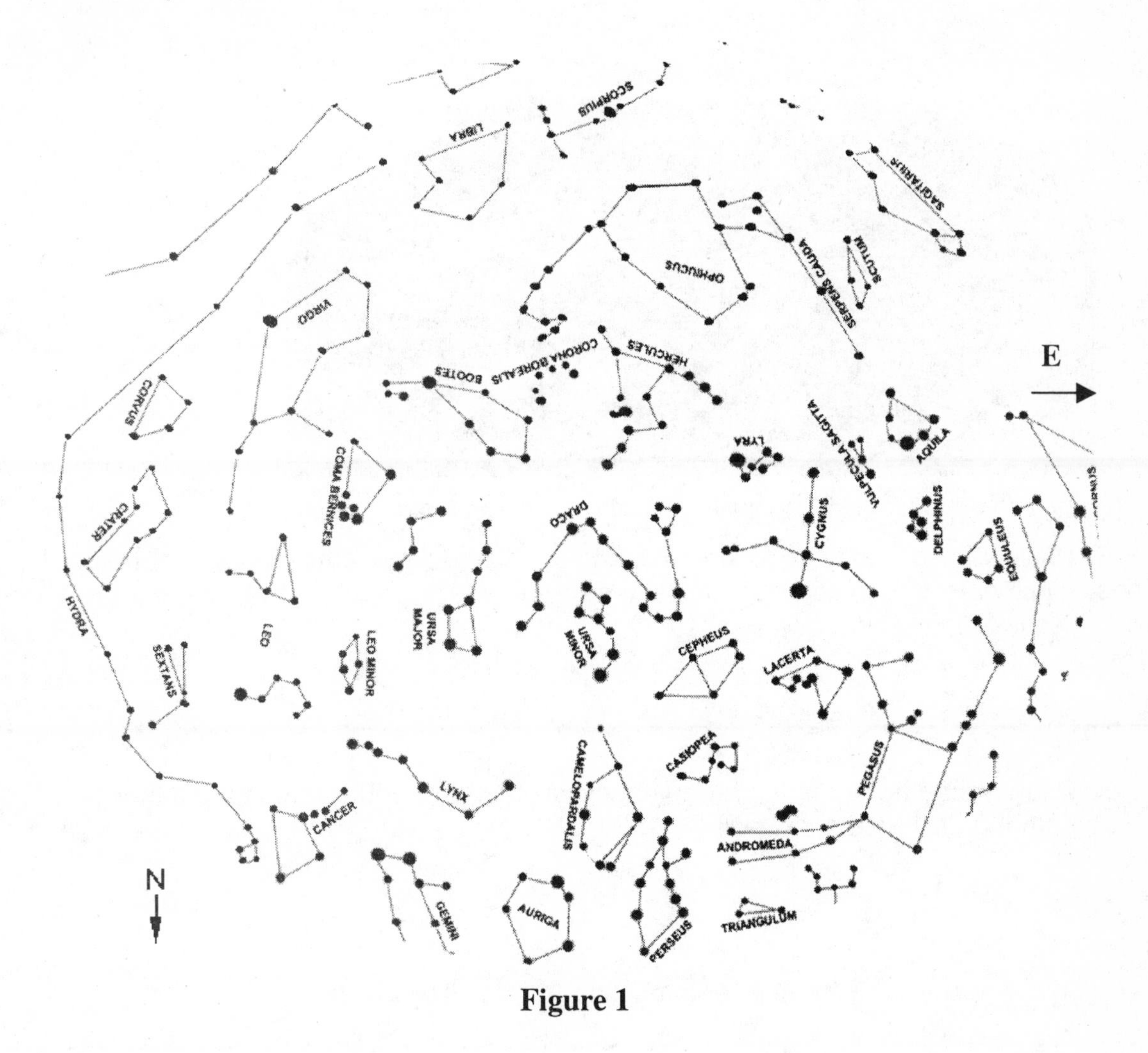

Figure 1

1) Which star group will appear highest in the night sky at this particular time?

2) Figure 2 shows a south-facing horizon view star map for July at midnight. What is the name of the star group that appears highest in the sky on this south-facing horizon view star map? (Hint: refer to the names provided in Figure 1.)

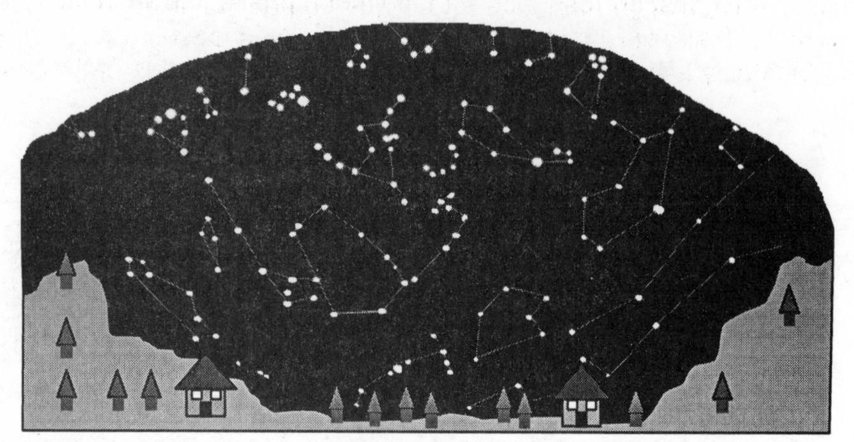

Figure 2

3) How would you have to hold, rotate, fold, and/or change the overhead view star map shown in Figure 1 so that it could be used as a south-facing star map like the one provided in Figure 2?

4) How would your answer to the previous question change if you wanted to use the star map from Figure 1 as a north-facing map?

5) Do you still agree with your answer to question 1? Why or why not?

6) When looking straight down at the overhead view star map from Figure 1:

a) where on the map is the star group that will appear highest in the night sky?

b) where on the map is the star group that will appear near the southern horizon?

c) where on the map is the star group that will appear near the eastern horizon?

Figure 1 shows Earth, the Sun, as well as five different possible positions for the Moon. It is important to recall that one half of the Moon's surface is illuminated by sunlight at all times. For each of the five small Moons shown below, the Moon has been shaded on one side to indicate the half of the Moon's surface that is not being illuminated by sunlight.

1) Which Moon position (A-E) best corresponds with the moon phase shown in the upper right corner of Figure 1? Make sure that the moon position you choose correctly predicts a moon phase in which only a small crescent of light on the left-hand side of the Moon is visible from Earth.

   Enter the letter of your choice:______

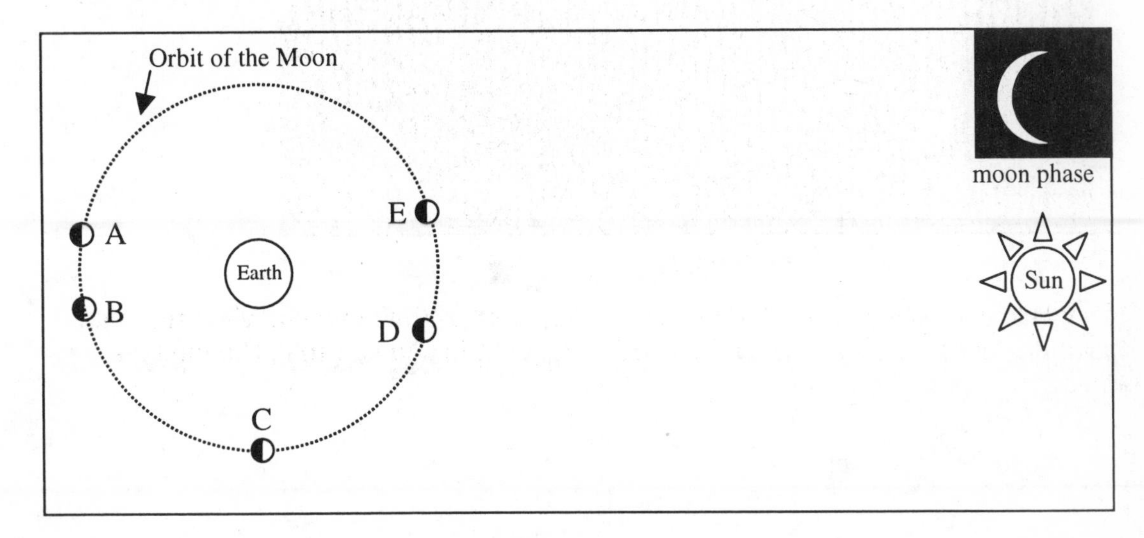

Figure 1

2) Give a reason why each of the other possible choices is not the correct position for the phase of the Moon shown in Figure 1.

3) Shade in each of the four Moons drawn in Figure 2 to indicate which portion of the Moon's surface will NOT be illuminated by sunlight.

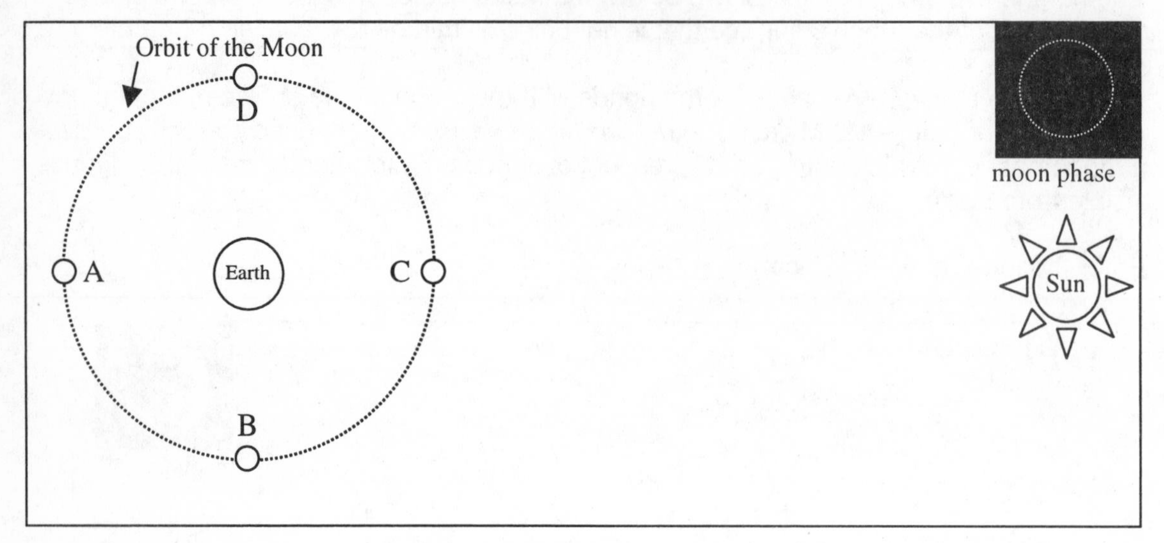

Figure 2

4) Which Moon position (A-D) best corresponds with the moon phase shown in the upper right corner of Figure 2?

Enter the letter of your choice:______

5) How much of the Moon's illuminated surface is visible from Earth for this phase of the Moon?

Circle One:

- None of the surface (visible from Earth) is illuminated
- Less than half of the surface (visible from Earth) is illuminated
- Half the surface (visible from Earth) is illuminated
- More than half the surface (visible from Earth) is illuminated
- All of the surface (visible from Earth) is illuminated

6) How much of the entire Moon's surface is illuminated by the Sun during this phase?

Circle One:

- None of the surface is illuminated
- Less than half of the surface is illuminated
- Half the surface is illuminated
- More than half the surface is illuminated
- All of the surface is illuminated

7) Is your answer to question 6 consistent with the information provided in the opening paragraph of this lecture tutorial? If so, explain how the two are consistent and if not, explain how they are different.

8) Consider the following debate between two students about the cause of the phases of the Moon.

Student 1: *The phase of the Moon depends on how the Moon, Sun and Earth are aligned with one another. During some alignments only a small portion of the Moon's surface will receive light from the sun, in which case we would see a crescent moon.*

Student 2: *I disagree. The moon would always get the same amount of sunlight; it's just that in some alignments Earth casts a larger shadow on the Moon. That's why the Moon isn't always a full moon.*

Do you agree or disagree with either or both students? Explain your reasoning for each.

1) Describe the differences between a waxing and a waning moon phase.

2) If the Moon is a full moon tonight, will the Moon be waxing or waning in one week? Which side of the Moon (right or left) will appear illuminated at this time?

   Circle One: Waxing or Waning

   Circle One: Right or Left

3) Where (in the southern sky, on the eastern horizon, on the western horizon, high in the sky, etc.) would you look to see the full moon when it starts to rise? What time of day would this happen?

4) If the Moon is a full moon when it rises in the evening, which of the phases shown below (A – H) will it be in when it sets? *Letter of Moon phase*: _________

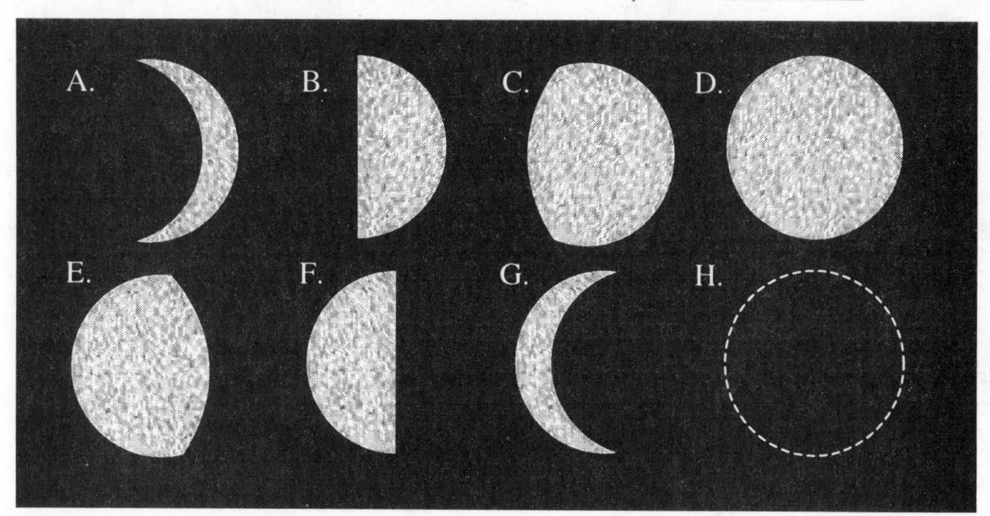

5) Where (in the southern sky, on the eastern horizon, on the western horizon, high in the sky, etc.) would you look to see the Sun when the full moon starts to rise?

6) Where (in the southern sky, on the eastern horizon, on the western horizon, high in the sky, etc.) would you look to see the new moon, if it were visible, when it starts to rise? What time of day would this happen?

7) Figure 1 shows the position of the Sun, Earth and Moon for a particular phase of the Moon. The Moon has been shaded on one side to indicate the portion of the Moon that is not being illuminated by sunlight. A stick figure person has been placed on the Earth to indicate an observer's position at noon. Note that in this representation the Earth will complete one counterclockwise rotation in each day.

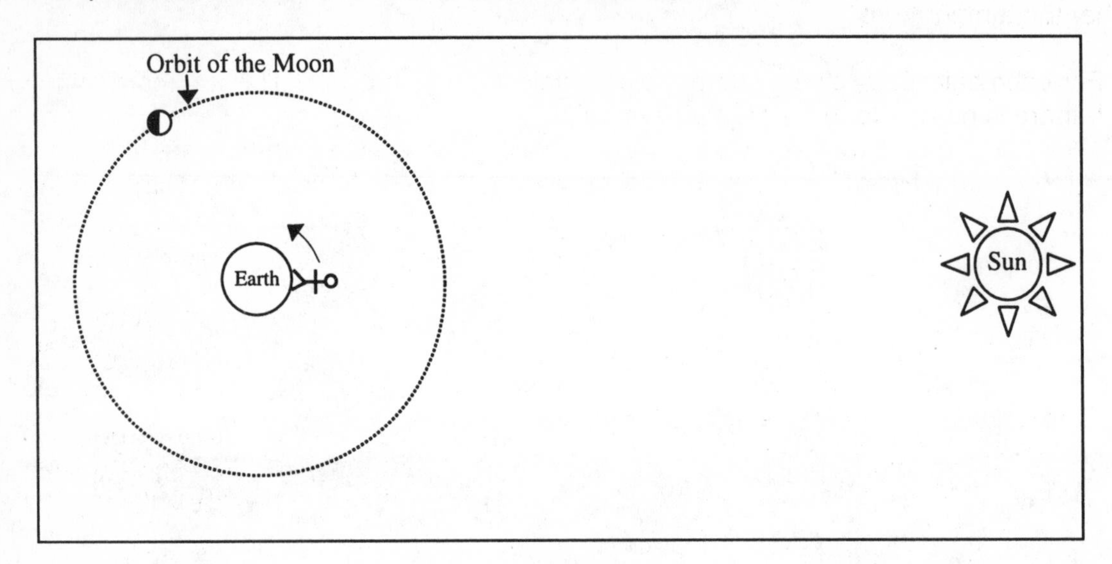

Figure 1

a) Which phase of the Moon is represented in Figure 1?

b) At what time will the Moon shown in Figure 1 appear highest in the sky (overhead)?

c) At what time will the Moon shown in Figure 1 rise?

d) At what time will the Moon shown in Figure 1 set?

8) If the Sun set below your western horizon about 2 hours ago, and the Moon is barely visible on the eastern horizon, is it a waxing or waning Moon that is visible?

9) A friend comments to you that he saw a beautiful, thin sliver of a Moon in the early morning on his way to work just before sunrise. Which phase of the Moon would this be, and in what direction would you look to see this moon phase (in the southern sky, on the eastern horizon, on the western horizon, high in the sky, etc.)?

Part I: Size, Temperature, and Luminosity

You are comparing the abilities of electric hot plates of different sizes and temperatures to bring identical pots of water to a boil. The pots are all as large as the largest hot plate. When a hot plate is at one of the temperature settings (low, med, high), the hot plate is depicted as a shade of gray as shown in question 1. The lighter the shade of gray, the higher the temperature.

1) For each pair of hot plates shown below, circle the one that will boil water more quickly. If there is no way to tell, state that explicitly.

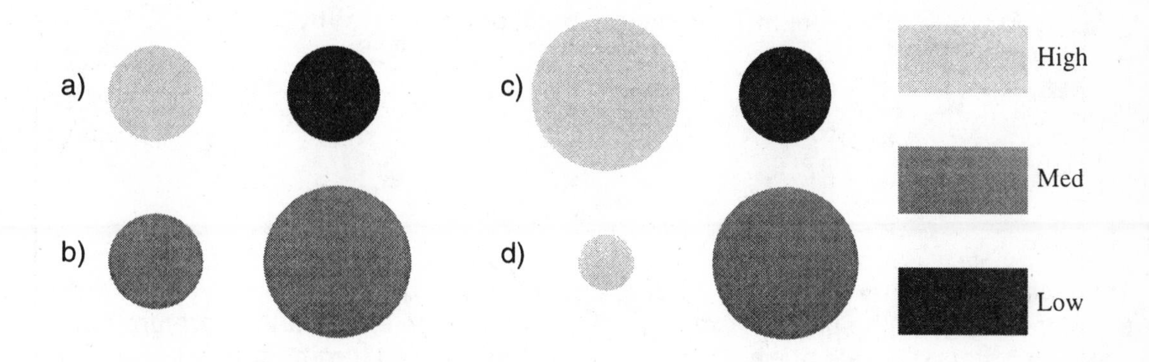

2) If you use two hot plates of the same size, can you assume that the one that boils water first is at the higher temperature? Which lettered example above supports your answer?

3) If you use two hot plates at the same temperature, can you assume that the one that boils water first is larger? Which lettered example above supports your answer?

4) If you use two hot plates of different sizes, can you assume that the one that boils water first is at a higher temperature? Which lettered example above supports your answer?

5) Two students are discussing their answers to question 4:

Student 1: *In 1d, the hot plate on the left boils water quicker than the one on the right even though it is smaller. The hot plate's higher temperature is what makes it boil water more quickly.*

Student 2: *But the size of the hot plate also plays a part in making it feel hot. If the hot plate on the left were the size of a pinhead, the water would take a long time to boil. I bet that if the size difference were great enough, the one at the lower temperature could boil the water first.*

Do you agree with Student 1, Student 2, both, or neither? Why?

The time for the water to come to a boil is determined by the rate at which the element transfers energy to the pot. This rate is related to both the size and temperature of the hot plate. For stars, the rate at which energy is emitted is called **luminosity**. Similar to the above example, a star's luminosity can be increased by:

- increasing its temperature; or
- increasing its surface area (or size).

This relationship between size, surface temperature, and luminosity is sometimes referred to as the **Stefan-Boltzmann law**. This law allows us to compare the sizes and temperatures of stars.

6) If two hot plates are at the same temperature and one boils water more quickly, what can you conclude about the sizes of the hot plates?

7) Likewise, if two stars have the same surface temperature and one is more luminous, what can you conclude about the sizes of the stars?

Part II: Application to the H-R Diagram

The graph below plots the luminosity of a star on the vertical axis against the star's surface temperature on the horizontal axis. This type of graph is called an H-R Diagram. Use the H-R Diagram below and the Stefan-Boltzmann law (as stated qualitatively above) in answering the following questions concerning the labeled stars.

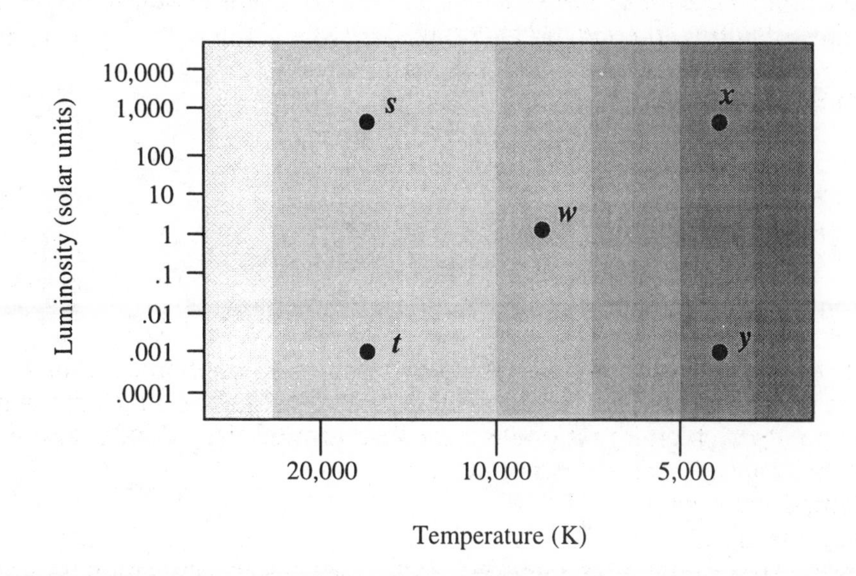

8) If two stars are at the same surface temperature and are the same size, which star, if either, is more luminous? Explain your answer.

9) Stars ***s*** and ***t*** have the same surface temperature. Given that Star ***s*** is actually much more luminous than Star ***t*** (i.e., it is giving off much more energy each second), what can you conclude about the size of Star ***s*** compared to Star ***t***? Explain your answer.

10) If two stars are the same size but one is at a higher surface temperature, which star, if either, is more luminous? Explain your answer.

11) Star ***s*** has a greater surface temperature than Star ***x***. Given that Star ***x*** is actually just as luminous as Star ***s***, what can you conclude about the size of Star ***x*** compared to Star ***s***? Explain your answer.

12) Based on the information presented in the H-R Diagram, which star is larger, ***x*** or ***y***? Explain.

13) Based on the information presented in the H-R Diagram, which star is larger, ***y*** or ***t***? Explain.

14) Why can't you compare the size of Star ***s*** to that of Star ***w***?

15) On the H-R Diagram, draw a "**z**" at the position of a star smaller in size than Star ***w*** but with the same luminosity. Explain your reasoning.

Part I: Spectral Curves

White light is made up of all colors of light. We can see the individual colors when we shine white light through a prism or look at a rainbow. The composition of light—called a spectrum—can be represented by a spectral curve, which shows the brightness of each color (or wavelength). Figure 1 shows a spectral curve for a source emitting more red and orange than indigo and violet. Notice that the red end of the curve is higher than the violet end so the object will appear slightly reddish in color. For a specific color on the horizontal axis, an object's brightness (or, more correctly, intensity) is represented by the height of the curve at that point.

1) Which color of light is most intense in Figure 1?

2) Imagine that the blue light and orange light from the source is now blocked. What color(s) are now present in the light?

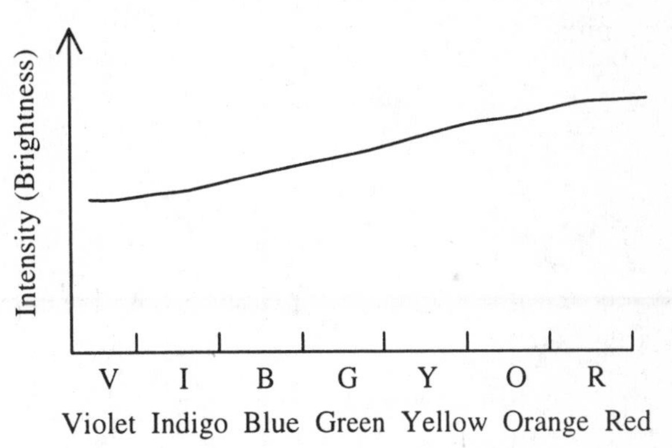

Figure 1

3) Which of the following is the most accurate spectral curve for the spectrum described in question 2?

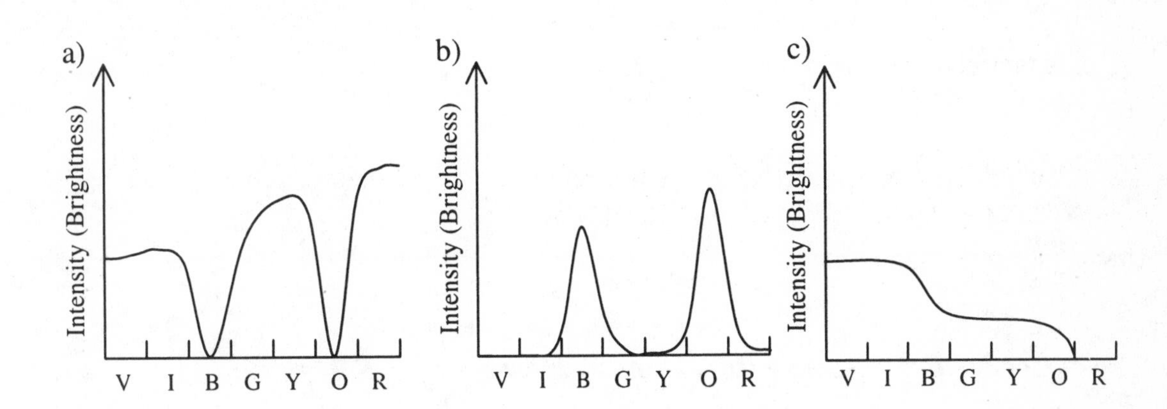

4) What colors of light are present in 3b above? Check your answer by discussing with another group.

5) What colors are present in 3c above? Would this appear reddish or bluish?

Part II: Blackbody Curves

Different colors of light are manifestations of the same phenomenon but have different wavelengths. For example, red light has a wavelength between 650nm and 750nm, while violet light has a wavelength between 350nm and 450nm. Stars also radiate light at wavelengths outside the visible range as seen in Figures 2a, 2b and 2c.

The two most important features of a star's blackbody radiation curve are its maximum height—an indication of the star's energy output—and the wavelength at which this occurs—called the peak wavelength. For example, if Star A and Star B are the same size and temperature, they will have identical blackbody curves. If Star B remains the same size but gets cooler, its output is reduced at all wavelengths and the peak occurs at a longer wavelength. This is shown in Figure 2a.

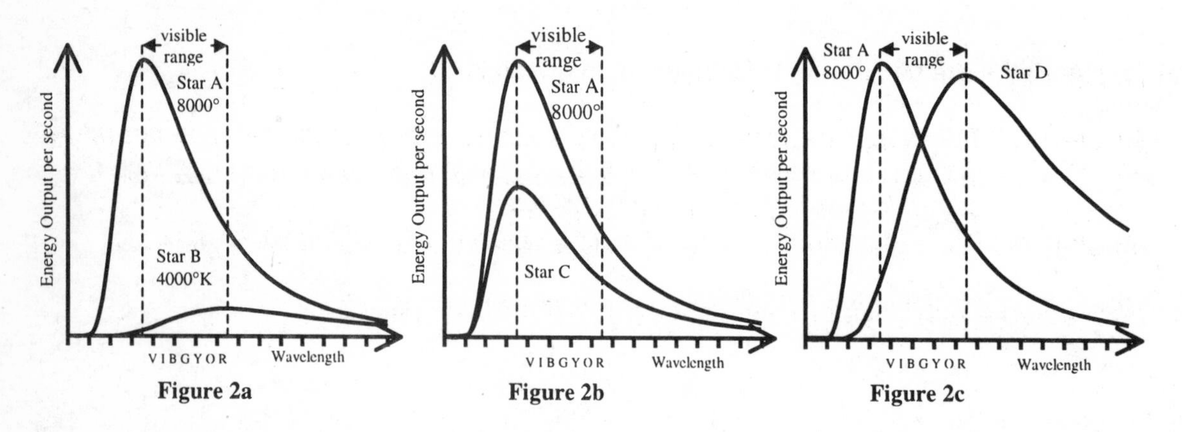

Figure 2a Figure 2b Figure 2c

Use Figure 2a to answer questions 6-9. Assume Stars A and B are the same size.

6) Which spectral curve has more red light?

7) Which spectral curve has more blue light?

8) Which star looks redder?

9) Two students are discussing their answers to question 8.

Student 1: *Star A looks redder because it is giving off more red light than Star B.*
Student 2: *But you're ignoring how much blue light Star A gives off. Star A gives off more blue light than red light so it looks bluish. Star B has more red than blue so it looks reddish. Star B is redder than Star A.*

Do you agree with Student 1, Student 2, both, or neither? Why?

10) Using the blackbody curves for the stars shown in Figure 2b, circle the correct answer for each characteristic of the curves below.

Longer peak wavelength	Star A	Star C	Same	
Lower surface temperature	Star A	Star C	Same	
Looks red	Star A	Star C	Both	Neither
Looks blue	Star A	Star C	Both	Neither
Greater energy output	Star A	Star C		

11) How must Star C be different from Star A to account for the difference in energy output?

12) Two students are discussing their answers to question 11.

Student 1: *The peaks are at the same place so they must be at the same temperature. If Star C were as big as Star A, it would have the same output. Since the output is lower, Star C must be smaller.*

Student 2: *No. If its output is lower, it must be cooler. It's still the same size.*

Do you agree with Student 1, Student 2, both, or neither? Why?

Consider the blackbody curves for the stars shown in Figure 2c.

13) For each star, describe its color as either reddish or bluish.
Star A: Star D:

14) Which star has the greater surface temperature? Explain your reasoning.

15) Which star is larger? Explain your reasoning. (Hint: one line of reasoning uses the fact that Stars A and B are the same size and Stars B and D are the same temperature.)

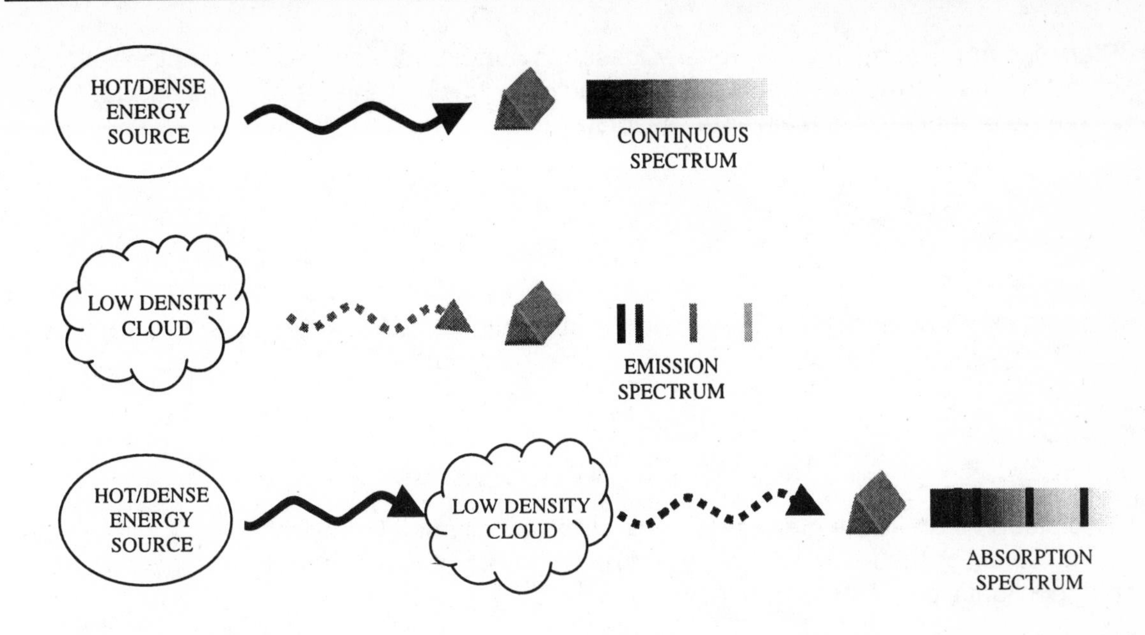

1) What type of spectrum is produced when the light emitted directly from a hot, dense object passes through a prism?

2) What type of spectrum is produced when the light emitted directly from a low density object passes through a prism?

3) Describe in detail the source of light and the path light must take to produce an absorption spectrum.

4) There are blank lines in the absorption spectum that represent missing light. What happened to the light that is missing from the absorption line spectrum?

5) Stars like our Sun have a low density, gaseous atmosphere surrounding their hot dense cores. If you were looking at the Sun's spectrum, which of the three types of spectrum would be produced? Explain your reasoning.

6) If a star existed that did NOT have a low density atmosphere surrounding it, what type of spectrum would you expect this particular star to produce?

7) Two students are looking at a brightly lit full moon, illuminated by reflected light from the Sun. Consider the following discussion between two students about what the spectrum of moonlight would look like.

Student 1: *I think moonlight is just reflected sunlight, so we will see the Sun's absorption line spectrum.*

Student 2: *I disagree, an absorption spectrum has to come from a hot dense object. Since the Moon is not a hot dense object it can't give off an absorption line spectrum.*

Do you agree or disagree with either or both of the students? Explain the reasoning behind your answer for each student.

8) Imagine that you are looking at two different spectra of the Sun. Spectrum #1 is obtained using a telescope that is in a high orbit far above Earth's atmosphere. Spectrum #2 is obtained using a telescope located on the surface of Earth. Label each spectrum below as either Spectrum #1 or Spectrum #2.

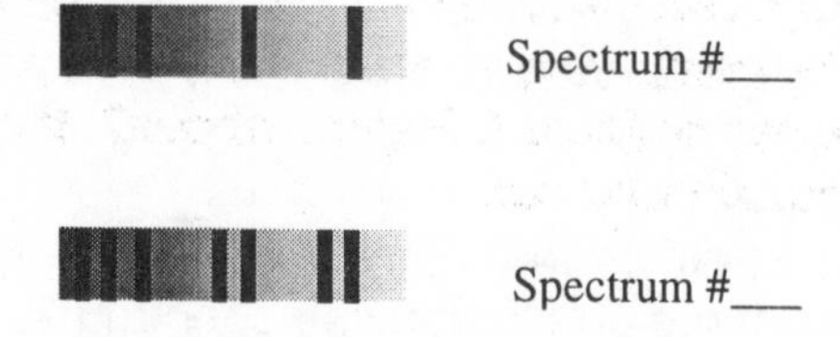

Explain the reasoning behind your choices.

The line absorption spectra for six hypothetical stars, each with different temperatures, are shown below. For each absorption line spectrum, the short wavelengths (or blue end) of the electromagnetic spectrum are shown on the left and the long wavelengths (or red end) of the spectrum are shown on the right.

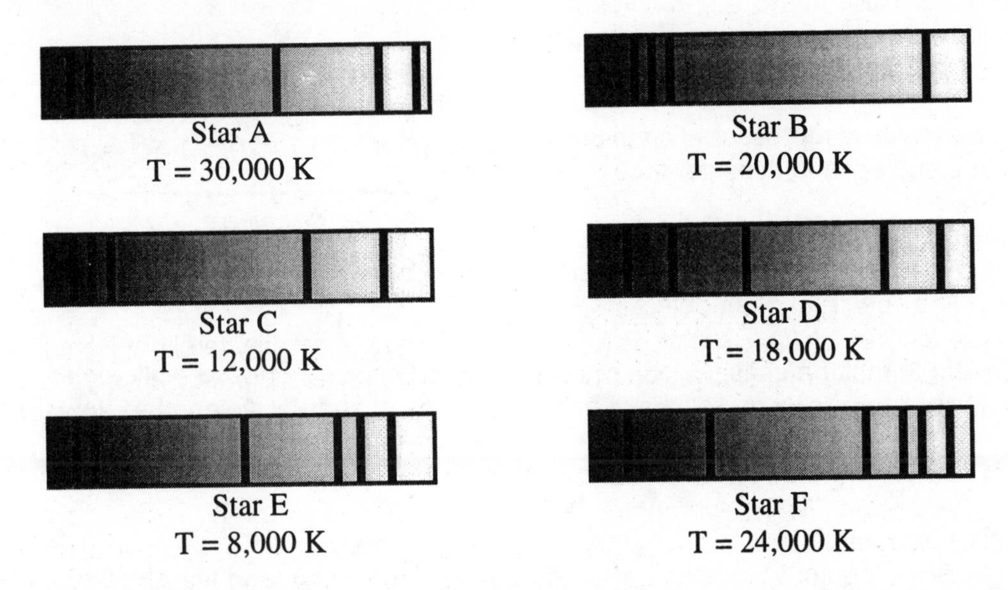

1) Do cold stars appear to have a different (greater or fewer) number of lines in their absorption spectra than hot stars? Cite evidence from the above spectra to support your answer.

2) Do cold stars always appear to have more lines at either the blue or red ends of their absorption spectra than hot stars? Cite evidence from the above spectra to support your answer.

3) Consider the absorption line spectrum given below for Star G. Can you determine the approximate temperature for Star G by comparing its absorption line spectrum to the absorption line spectra and temperatures of Stars A-F given above? If so, write in your estimate in the space below; if not, explain why not.

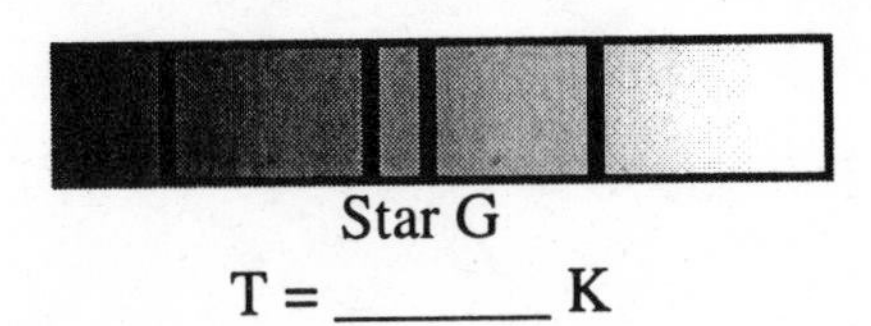

The spectral curve on the graph at right illustrates the energy output versus wavelength for Star G. Again, the short wavelengths (or blue end) of the electromagnetic spectrum are represented on the left side of the horizontal axis and the long wavelengths (or red end) of the spectrum are represented on the right end of the horizontal axis.

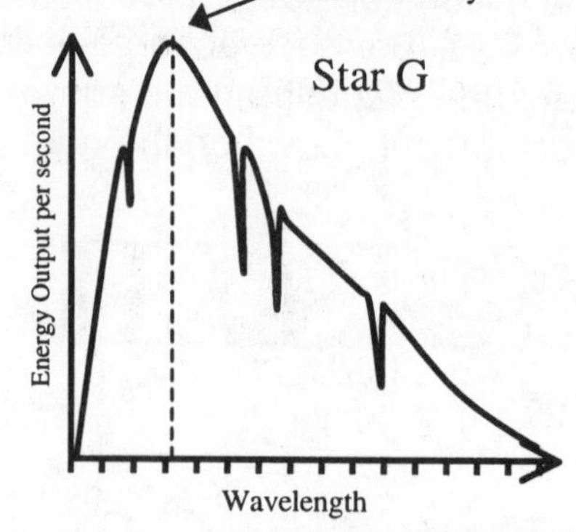

The two important features represented on this spectral curve that you need to consider are:

a) The exact locations of the small dips, or "absorption features," on the curve correspond to the dark lines that appear in the absorption line spectrum of an object.

b) The wavelength that matches the object's peak intensity corresponds directly to the object's temperature. Hot objects have peak intensities at short wavelengths—toward the blue end of the spectrum. Cooler objects have peak intensities at long wavelengths—toward the red end of the spectrum.

Although the total energy output of a star is affected by its temperature (and therefore so is the height of the spectral curve), for this activity, we will assume that the height (but not location) of the peak intensity and that the general shape of the spectral curves for Stars A – F can be drawn nearly the same for each star. Only the location of the peak intensity and the location of the small dips, or absorption features, will be different for each of the stars.

4) Examine the spectral curve shown at right for Star A. Star G has a temperature of approximately 25,000 K. Is the location for the peak intensity shown in the sketch of the spectral curve for Star A drawn at approximately the correct wavelength? Explain your reasoning.

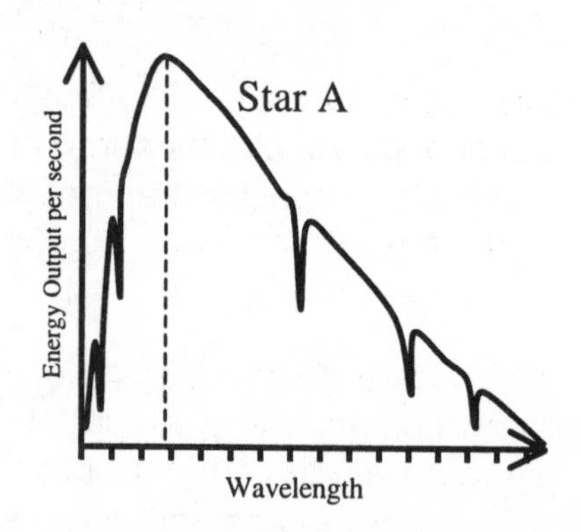

5) Are the absorption features (dips) in the spectral curve for Star A drawn at approximately the correct wavelengths? Explain how you can tell.

6) Sketch spectral curves for Stars B-F on the corresponding graphs provided below. Make sure the absorption features and the peak intensities are located at approximately the correct wavelengths.

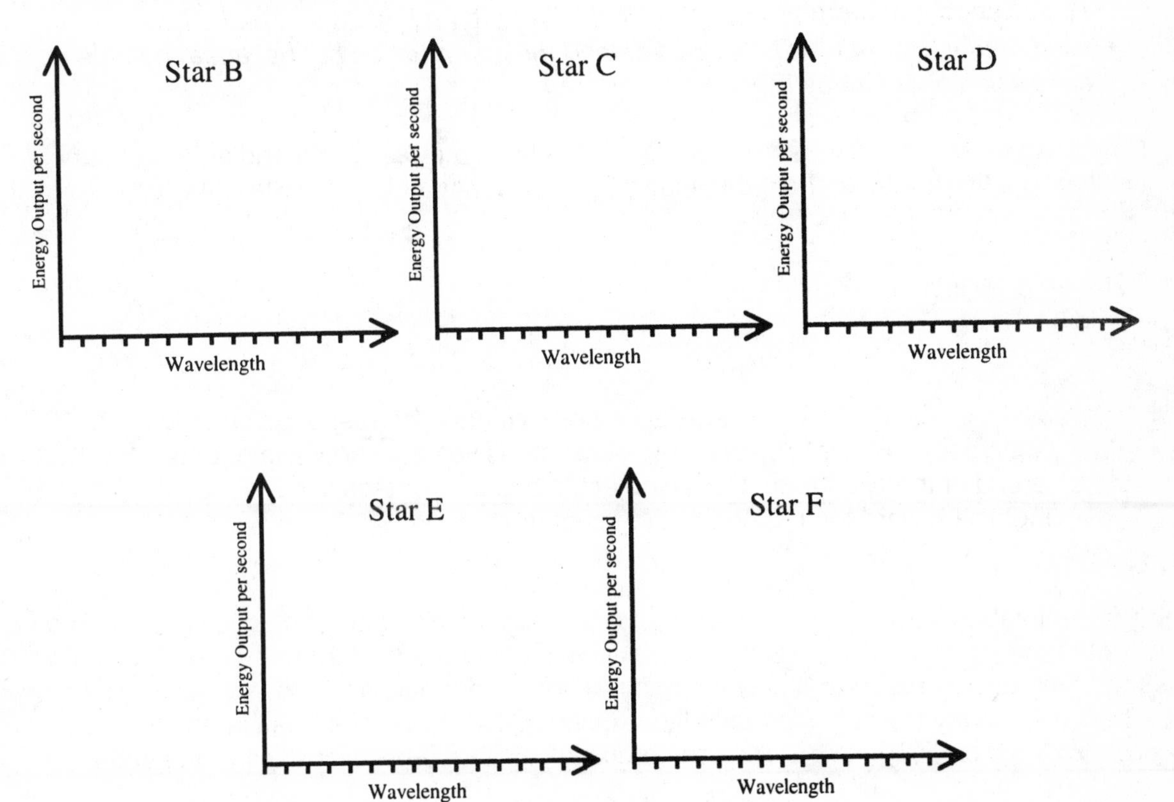

7) Did you draw the peak intensity of each spectral curve at the same wavelength as the spectral curve for Star A? Why or why not?

8) If you were given a star's absorption line spectrum and its corresponding spectral curve shown on an energy output per second versus wavelength graph, how could you approximate the temperature of the star?

9) Consider the following statement made by a student regarding a star's temperature and its corresponding absorption line spectrum.

If I am looking at a star's absorption line spectrum and see that it has a lot of lines at the blue end of the spectrum, then the star must be hot because the blue lines are higher energy lines.

Do you agree or disagree with this student? Explain your reasoning and support your answer by citing evidence from the absorption line spectra given for Stars A – G.

The drawing shown below illustrates the amount that different wavelengths of light are able to penetrate Earth's atmosphere. The dotted/shaded regions are used in this drawing to depict the different layers and densities of air in Earth's atmosphere. Notice that the atmosphere can be transparent to light at some wavelengths (lines passing through the atmosphere to the surface of Earth) and yet can also completely block other wavelengths of light (lines being absorbed in the atmosphere).

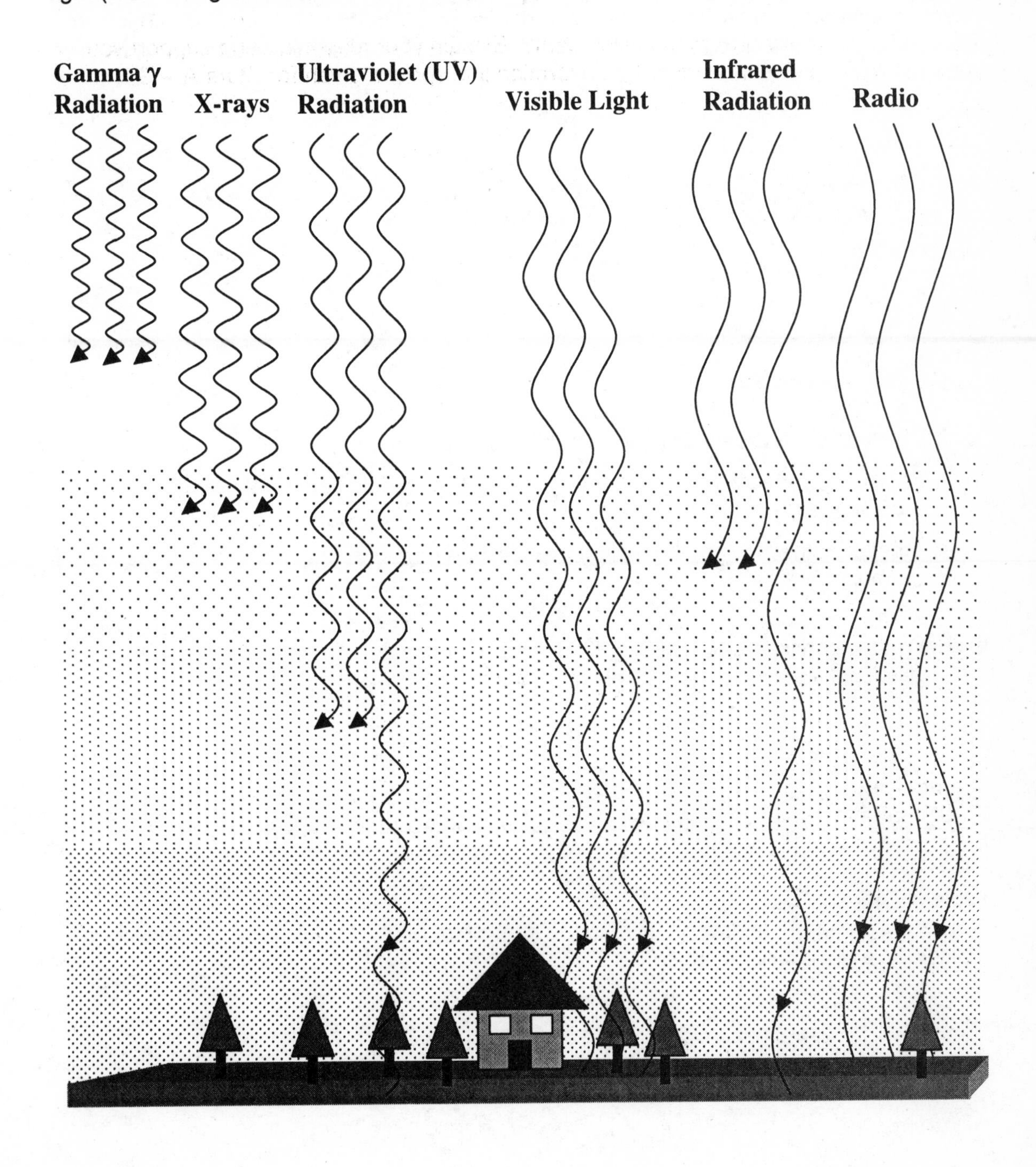

1) Which, if any, of the different wavelengths of light (electromagnetic radiation) shown in the above image are able to **entirely** penetrate Earth's atmosphere and reach the surface?

2) Which, if any, of the different wavelengths of light (electromagnetic radiation) shown in the above image are able to **partially** penetrate Earth's atmosphere and reach the surface?

3) Which, if any, of the different wavelengths of light (electromagnetic radiation) shown in the above image are **not at all** able to reach Earth's surface?

4) Federal funding agencies must form committees to decide which telescope projects will receive funds for construction. When deciding which projects will be funded, the committees must consider:

   i. how efficiently telescopes work at different wavelengths,

   ii. that telescopes in space are much more expensive to construct than Earth-based telescopes, and

   iii. that certain wavelengths of light are blocked from reaching Earth's surface by the atmosphere.

   Knowing this, consider each pairing of telescope proposals listed below (a-d) and state which proposal you would fund. Explain the reasoning behind your decisions.

   a)

   Project Delta:
   A gamma ray wavelength telescope, located in Antarctica, used to look for evidence of the presence of a black hole.

   Project Theta:
   A visible wavelength telescope used in the search for planets outside the solar system, located on a university campus.

b)

Project Beta:
An X-ray wavelength telescope, designed specifically to look at the Sun, that is located near the North Pole.

Project Alpha:
An infrared wavelength telescope used to view supernovae, placed on a satellite in orbit around Earth.

c)

Project Rho:
A UV wavelength telescope used to look at distant galaxies and placed high atop Mauna Kea in Hawaii at 14,000 ft above sea level.

Project Sigma:
A visible wavelength telescope used to observe a pair of binary stars located in the constellation Ursa Major, placed on a satellite in orbit around Earth.

d)

Project Zeta:
A radio wavelength telescope that is placed on the floor of the Mojave desert, used to detect potential communications from distant civilizations outside our solar system.

Project Epsilon:
An infrared wavelength telescope, located in the high elevation mountains of Chile, used to view newly forming stars (protostars) in the Orion nebula.

The most difficult part of constructing an accurate model for planetary motions is that planets seem to wander among the stars. During their normal motion, planets appear to move from west to east over many consecutive nights as seen against the background stars. However, they occasionally (and predictably) appear to reverse direction and move east to west over consecutive nights. This backwards motion is called retrograde motion.

1) Given the data in Table 1, plot the motion of the mystery planet on the graph provided in Figure 1 (record dates next to each position you plot). Then, draw a smooth line, using your data points, to illustrate the path of the planet through the sky.

Date of Observation	Azimuth Direction (horizontal)	Altitude Direction (vertical)
May 1	240	45
May 15	210	55
June 1	170	50
June 15	150	40
July 1	170	35
July 15	180	45
August 1	140	50
August 15	120	55

Table 1 – Mystery Planet Positions

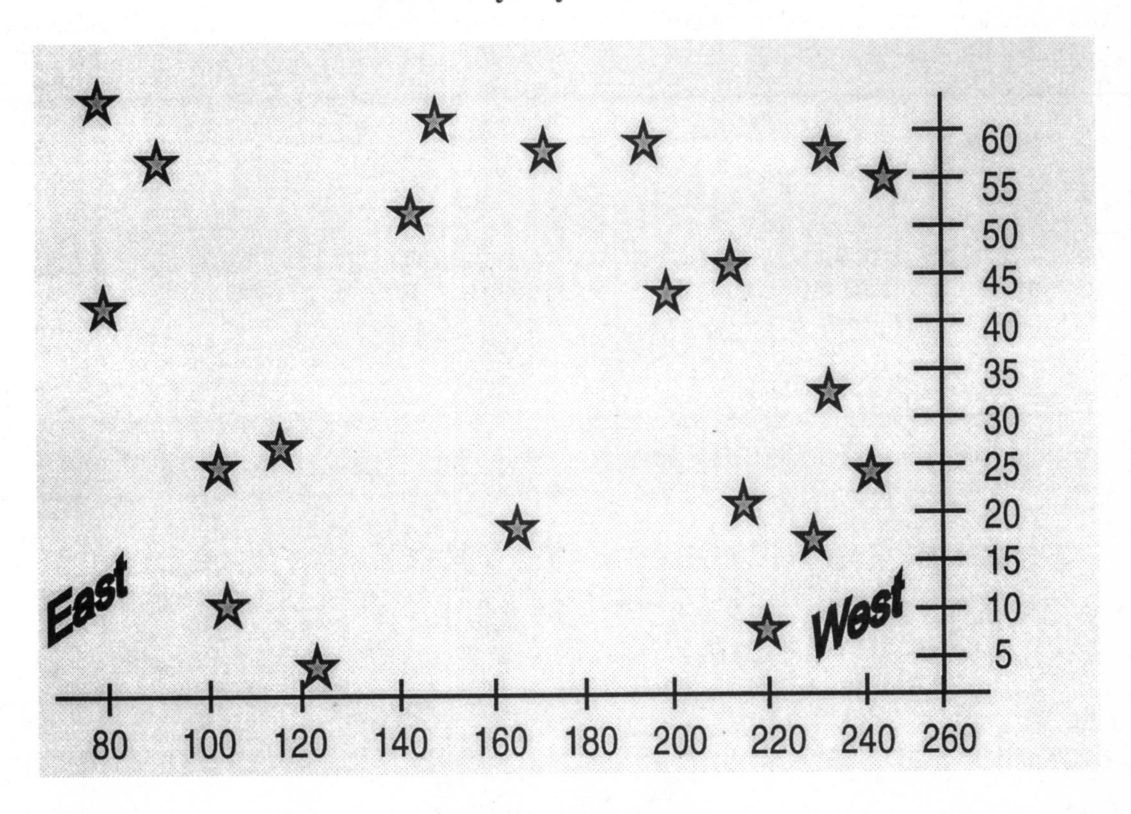

Figure 1 – Planet Path

2) On what date was the mystery planet located farthest to the West? What was the Azimuth direction of the planet on this date?

3) On what date was the mystery planet located farthest to the East? What was the Azimuth direction of the planet on this date?

4) Describe how the mystery planet moved (east or west), as compared to the background stars, during the time between the dates identified in questions 2 and 3.

5) During which dates does the mystery planet appear to move with normal motion as compared to the background stars? In what direction does the planet appear to be moving relative to the background stars during this time?

6) During which dates does this mystery planet appear to move with retrograde motion as compared to the background stars? In what direction does the planet appear to be moving relative to the background stars during this time?

7) If a planet were moving with retrograde motion, how would the planet appear to move across the sky in a single night? Where would it rise? Where would it set?

8) Suppose your instructor says that Mars is moving with retrograde motion tonight and will rise at midnight. Consider the following student statement:

    Since Mars is moving with retrograde motion that means that during the night it will be moving West to East rather than East to West. So at midnight it will rise in the West and move across the sky and then later set in the East.

    State whether you agree or disagree with this student's statement. Explain the reasoning behind your decision.

Orbital Period and Orbital Distance

In this lecture tutorial we will investigate the relationship between how long it takes a planet to orbit the Sun (orbital period) and how far away a planet is from the Sun (orbital distance).

1) Consider an imaginary solar system in which a large Jupiter-sized planet is in orbit close to a Sun-sized star, while a small Mars-sized planet is in orbit around the same star but at a greater distance. Which of the two planets do you think will have the shorter orbital period? Why?

2) If the large Jupiter-sized planet and the small Mars-sized planet were to switch positions, would your answer to question 1 change? If so, how? If not, why not?

3) Do you think that the orbital period for the Jupiter-sized planet would increase, decrease or stay the same if it were to move from being very close to the central star to being much farther from the star? Why?

4) Imagine the large Jupiter-sized planet and the small Mars-sized planet were able to orbit the central star at the same distance and never collide. Do you think the two planets would have the same or different orbital periods? Why?

The graph below illustrates how the orbital period and orbital distance of a planet are related.

5) Does a planet's orbital period appear to increase, decrease or stay the same as a planet's orbital distance is increased?

6) How far from the central star does a planet orbit if it has an orbital period of 1 year?

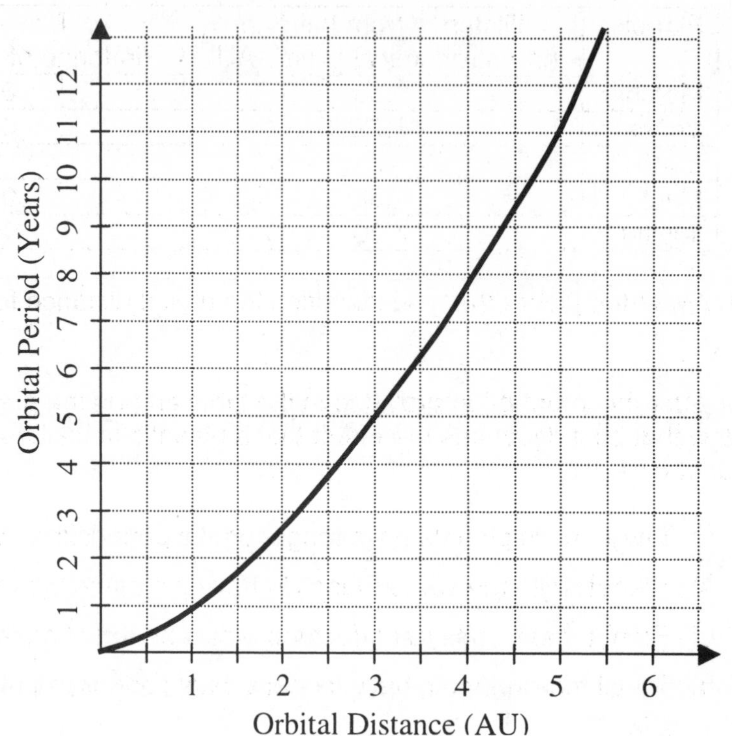

7) How long does it take a planet to complete one orbit if it is twice the distance from the central star as the planet from question 6?

8) Based on your results from questions 6 and 7 which of the following best describes the relationship between a planet's orbital distance and orbital period? Circle your choice.

   a) Doubling a planet's orbital distance causes its orbital period to decrease by half.
   b) Doubling a planet's orbital distance causes its orbital period to double.
   c) Doubling a planet's orbital distance causes its orbital period to more than double.
   d) Doubling a planet's orbital distance does not change its orbital period.

In the table below we have provided the orbital distances and masses for the closest five planets to the Sun.

Planet	Distance from the Sun (in Astronomical Units - AU)	Planet mass (in terms of Earth's mass)
Mercury	0.38	0.06
Venus	0.72	0.82
Earth	1.0	1.0
Mars	1.52	0.11
Jupiter	5.20	318

9) What was the planet that you identified the orbital distance for in question 6?

10) Consider the information provided in the table and on the graph and choose the answer below that best describes the effect that a planet's mass has on its orbital period. Circle your choice.

a) Large mass planets have longer orbital periods than small mass planets.

b) Planets with the same mass will have the same orbital period.

c) Planet mass does not affect the orbital period of a planet.

d) Small mass planets have longer orbital periods than large mass planets.

Explain the reasoning behind your choice.

11) Review your answers to questions 1-4. Do you still agree with the answers you provided? If not, describe (next to your original answers) how you would change the answers you gave initially.

The extremely high temperature of Earth's core causes material in the surrounding mantle to become hot, to expand, and to rise toward the surface. The mantle material then cools and sinks, resulting in a circular motion of material moving beneath Earth's surface. This circulation of mantle material causes the continental and oceanic plates to move across Earth's surface. At various locations on Earth's surface, we are able to observe plates colliding, plates separating, and plates moving horizontally.

The drawing below shows a cross section of Earth's surface and its underlying mantle. At this particular location of the surface, the dense oceanic plate is being forced beneath the less dense continental crust. The dense oceanic plate experiences higher temperatures (and pressures) as it is forced deeper into the mantle. This interaction between the oceanic plate and continental plate causes molten material to move upward through the continental plate until it breaks the surface in the form of volcanoes.

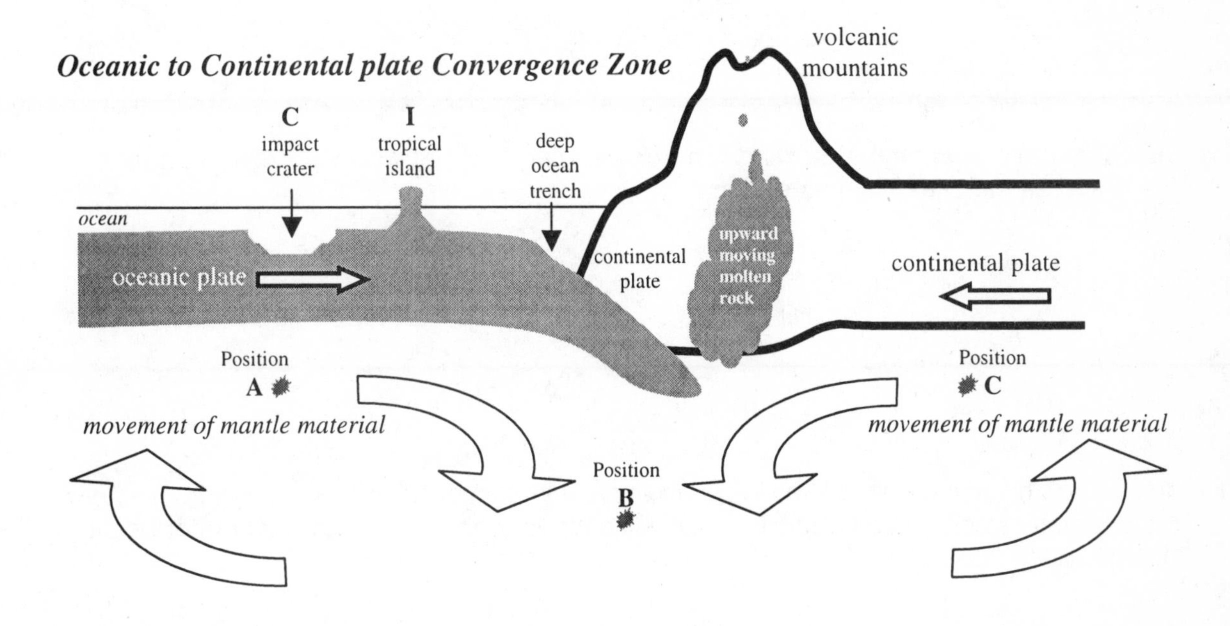

1) In the drawing above, sketch arrows showing the direction mantle material is moving at positions A, B, and C.

2) In the drawing above, which way (*right or left*) are the oceanic and continental plates moving?

3) Consider the following statements made by two students debating why the oceanic and continental plates move.

Student 1: *The plates are moving because the mantle material is constantly moving beneath Earth's plates, and this causes the plates to move.*

Student 2: *I disagree. The plates are just floating on the mantle material. The plates started moving a long time ago when Earth initially formed and the plate's inertia keeps them moving toward each other.*

Do you agree with either or both students? Why?

4) Just beneath *point I* on the drawing is a tropical island. What will eventually happen to the island as the oceanic plate moves?

5) Just beneath *point C* on the drawing is an ancient impact crater on the ocean floor where a giant meteor collided with Earth. What will happen to the ancient impact crater as the oceanic plate moves?

6) Imagine that an impact occurred on the continental plate millions of years ago that left behind an impact crater near the base of a volcano. Why would there be little evidence of this impact crater found today?

PRELIMINARY EDITION, 2002

7) Consider the image below of our rocky and crater-covered Moon. This very old surface has remained virtually unchanged over the last few billions of years. What can you infer about the interior of our Moon? Explain your reasoning.

8) If a new planet were discovered, what would you look for to determine whether or not it has a hot and molten interior? Why?

Temperature and Formation of Our Solar System

Consider the information provided in the graph and table below. The graph indicates the temperature at different locations in the solar system during the time when the planets were forming as part of the solar nebula. The table provides some common temperatures to use for comparison.

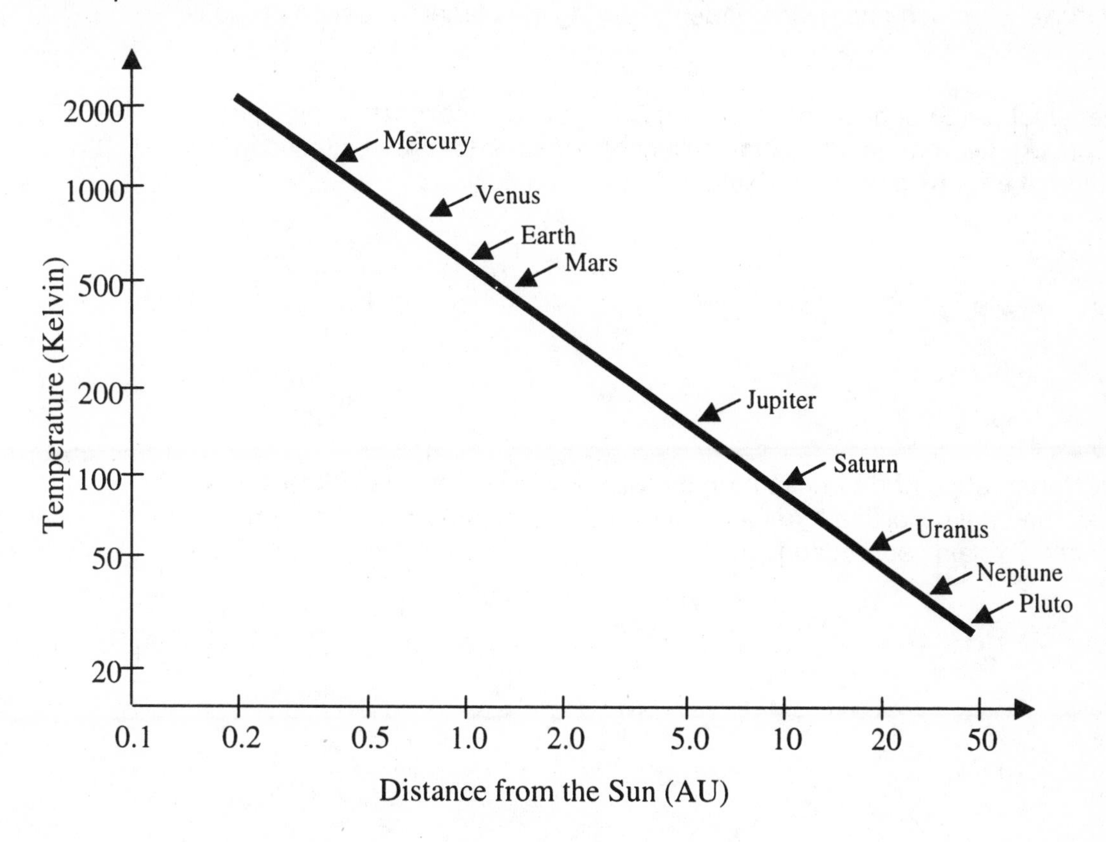

Condition	Temp. Fahrenheit	Temp. Celsius	Temp. Kelvin
Severe Earth Cold	-100	- 73	199
Water Freezes	32	0	273
Room Temp	72	22	296
Human Body	98.6	37	310
Water Boils	212	100	373

1) What was the temperature in the solar nebula at the location of Earth?

2) What was the temperature in the solar nebula at the location of the asteroid belt (2.8 AU)?

3) Which planets formed at temperatures lower than the freezing point of water?

4) Which planets formed at temperatures greater than the freezing point of water?

At temperatures greater than the freezing point of water, light gases, like hydrogen and helium, have too much energy to condense during the planet formation process.

5) Over what range of distances from the Sun would you expect to find light gases, like hydrogen and helium, collecting together to form a significant portion of a forming planet? Explain your reasoning.

6) Over what range of distances from the Sun would you expect to find solid, rocky material collecting together to form a significant portion of a forming planet? Explain your reasoning.

7) If a planet had formed at 2.2 AU from the Sun, what type of planet (terrestrial or Jovian) would you expect to find? Explain your reasoning.

8) Could a large Jovian planet have formed at the location of Mercury? Explain your reasoning.

PART I: Earth and Moon

Astronomers often deal with large numbers for distances, masses, and other quantities. They often use ratios to get a better sense of how big or small these quantities are. This can be useful in our daily life as well. For example, we may not have a good sense for the length of a 40-meter-long commercial jet, but saying that the jet is 8 times longer than a car may be more meaningful to us.

Distances like the following can be hard to conceptualize:

Moon's diameter: 3,476 km
Earth's diameter: 12,756 km

But we can think about these sizes in terms of one another so that we can create a scale model of Earth and the Moon in our minds. If we say, "Earth is (some number) times bigger than the Moon," we can determine that number by dividing Earth's diameter by the Moon's diameter. The result is roughly 4 (12,756/3476 ≈ 4). This means it would take about four Moons to stretch across the diameter of Earth (as shown below).

1) Which of the following pairs of objects would make a good scale model of Earth and the Moon?

   a) A basketball and a soccer ball
   b) A basketball and a softball
   c) A basketball and a ping pong ball
   d) A basketball and a pea
   e) A basketball and a grain of sand

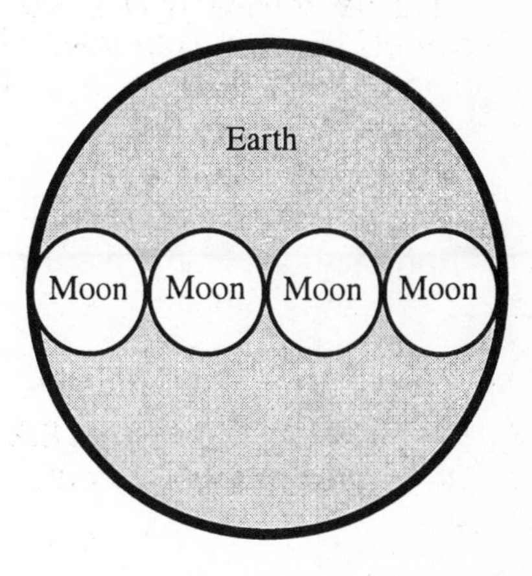

2) Could any combination of the "second" objects (soccer ball, softball, ping pong ball, pea, grain of sand) also be used to make a scale model of Earth and the Moon?

The distance between Earth and the Moon is much larger than either the Moon or Earth – but how much larger? If we divide the Earth-Moon distance (384,000 km) by Earth's diameter, we get 384,400/12,756 ≈ 30. This means you could fit about 30 Earths between Earth and the Moon.

3) To make a scale model of the Earth-Moon orbital system, you not only need to pick appropriately sized objects to represent Earth and the Moon; you also need to place them the right distance apart. Let's say you use a 12-inch (1 foot) basketball and a 3-inch orange as your Earth and Moon respectively. About how far apart must they be to have an accurate scale model of the Earth-Moon orbital system? (circle your answer below) Why?

   a) 1 foot
   b) 4 feet
   c) 10 feet
   d) 30 feet
   e) 300 feet

PART II: The Sun

Compared to the size of Earth, the Sun (with diameter 1,392,000 km) is about 110 times bigger than Earth.

4) Can any combinations of the following items be used to make accurate scale models of Earth and the Sun? If so, which two would you choose and why? If not, why not?
   - basketball
   - soccer ball
   - softball
   - ping pong ball
   - pea
   - grain of sand

Now let's compare the Sun's diameter to the size of the Moon's orbit around Earth. The diameter of the Moon's orbit is about 769,000 km across. So, the ratio of the Sun's diameter to the Moon's orbital diameter is roughly 2. (1,392,000 / 769,000 ≈ 2)

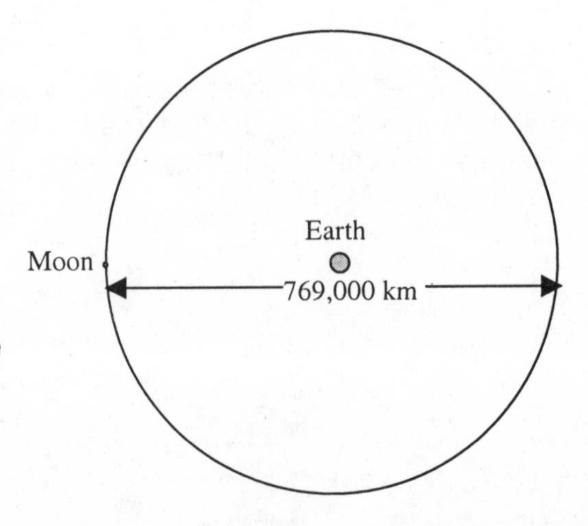

5) Does this mean that two Suns placed side by side would fit inside the Moon's orbit or that the two Moon orbits placed side by side would fit across the Sun? Explain your answer.

The distance from Earth to the Sun is about 150,000,000 km. This makes the Sun-Earth distance about 110 times larger than the size of the Sun (150,000,000/1,392,000 ≈ 110).

6) If we used a 12-inch (1 foot) basketball to represent the Sun, how far would it have to be from Earth to be an accurate scale model?
   a) 1 foot
   b) 10 feet
   c) 30 feet
   d) 110 feet
   e) 300 feet

7) If we used a basketball to represent the Sun and a ping-pong ball to represent Earth, and separated them by the distance you answered in question 6, would we have an accurate scale model of the Sun-Earth system? Explain your answer.

8) How many Moons would fit across the diameter of the Sun?

1) Which value, apparent magnitude or absolute magnitude, do you think:

   a) tells us how bright an object will appear from Earth?

   b) tells us about the object's actual brightness?

2) Consider the following debate between two students.

   Student 1: *I think that stars with the greatest actual brightness also have a large apparent magnitude and a large absolute magnitude.*
   Student 2: *I don't think that apparent or absolute magnitudes have anything to do with it, because they are not related to the brightness of a star. The brightness of a star just depends on how much light the star gives off.*

   Do you agree or disagree with either or both of the student statements? Explain your reasoning for each.

3) Star Y appears much brighter than Star Z when viewed from Earth, but is found to actually give off much less light. Assign a set of possible values for the apparent and absolute magnitudes of these stars that would be consistent with the information given in the previous statement.

4) The star Rigel has an apparent magnitude of 0.1 and is located about 250 parsecs away from Earth. Which of the following is the most likely absolute magnitude for Rigel?

   a) -6.9
   b) 0.1
   c) 7.1

   Explain your answer.

5) Refer to the following table for questions 5(a)-(d):

	apparent magnitude	*absolute magnitude*
Star A:	1	1
Star B:	1	2
Star C:	5	4
Star D:	4	4

a) Which object appears brighter from Earth: Star C or Star D? Explain your reasoning.

b) Which object is actually brighter: Star A or Star D? Explain your reasoning.

c) How would the apparent and absolute magnitudes of Star A change if it were located at a distance of 40 parsecs? Explain your reasoning.

d) Rank the objects (from farthest to closest) in order of their distances from Earth. Explain your reasoning for your ranking.

6) Star F is known to have an apparent magnitude of –26 and an absolute magnitude of 4. Where might this star be located? What is the name of this star? Explain your reasoning.

Part I: Stars in the sky

Consider the diagram to the right:

1) Imagine that you are looking at the stars from Earth in January. Use a straightedge or a ruler to determine which of the distant stars appears closest to Star A in your night sky. Circle this star and label it "Jan."

2) Repeat question 1 for July and label the star "July."

3) Below, these same distant stars are shown as you would see them in the night sky. In this box, draw a small x to indicate the position of Star A as seen in January and label it "Jan."

4) In the same box, draw another x to indicate the position of Star A as seen in July and label it "July."

5) Describe the motion of Star A relative to the distant stars as Earth orbits the Sun counterclockwise from January of one year, through July, to January of the following year.

Nearby star
(Star A)

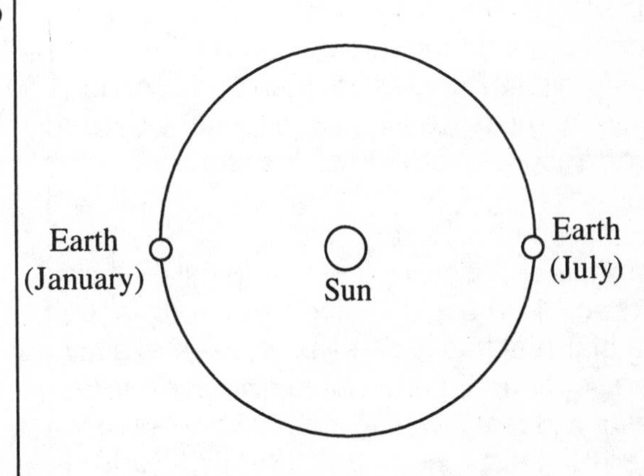

The apparent motion of nearby objects relative to distant objects, which you just described, is called **parallax**.

6) Consider two stars that both exhibit parallax. Star C appears to move back and forth more than Star D. Which star is closer (circle one)? If you're not sure, just take a guess. We'll return to this question later.

   Star C   Star D

Part II: What's a parsec?

Consider the diagram to the right:

7) Starting from Earth in January, draw a line through Star A to the top of the page.

8) There is now a narrow triangle with the Earth-Sun distance as its base. The small angle, just below Star A, formed by the two longest sides of this triangle is called the **parallax angle** for Star A. Label this angle "$\mathbf{p_A}$."

Knowing a star's parallax angle allows us to calculate the distance to the star. Because even the nearest stars are still very far away, parallax angles are extremely small. They are measured in "arcseconds" where an arcsecond is 1/3600 of 1 degree.

To describe the distances to stars, astronomers use a unit of length called the **parsec**. 1 parsec is defined as the distance to a star that has a **par**allax angle of exactly 1 arc**sec**ond. The distance from the Sun to a star 1 parsec away is 206,265 times the Earth-Sun distance or 206,265 AU. (Note that the diagram to the right is not drawn to scale.)

9) If the parallax angle (for Star A) $\mathbf{p_A}$ is 1 arcsecond, what is distance from the Sun to Star A? (Hint: use parsec as your unit of distance.) Label this distance on the diagram.

10) Is a parsec a unit of length or a unit of angle? It can't be both.

Important Note: The distance from the Sun to even the closest star is so much greater than 1 AU that we always consider the distance from Earth to a star and the distance from the Sun to that star to be the same.

Part III: Distances

11) Consider the following discussion between two students working on this tutorial regarding the relationship between parallax angle and the distance we measure to a star:

Student 1: *If the distance to the star is more than 1 parsec, then the parallax angle must be more than 1 arcsecond. Larger distance means larger angle.*

Student 2: *If we drew a diagram for a star that was much more than 1 parsec away from us, the triangle in the diagram would be pointier than the one we just drew in Part II. That should make the parallax angle smaller for a star farther away.*

Do you agree with Student 1, Student 2, both, or neither? Why?

12) On your diagram from Part II, draw a second star on the dotted line farther from the Sun than Star A. Do this in another color, if possible, and label it "Star B." Repeat steps 7 and 8 from Part II, except label the parallax angle $\mathbf{p_B}$.

13) Which star, the closer one or the farther one, has the larger parallax angle? Check your answer against your answers for questions 6 and 11 and resolve any discrepancies.

Part I: Angular Measurement

Imagine that you are standing in an open field. Looking due south, you see a house in the distance. If you look due east, you see a barn in the distance.

1) What is the angle between the house and the barn? (Hint: If you point at the barn with one arm and point at the house with your other arm, what angle do your arms make?)

2) You see the Moon on the horizon just above the barn in the east, and also see a bright star just above the house in the south. What is the angle between the Moon and the bright star?

3) Compare your answers for the barn-house angle from question 1 and the Moon-star angle from question 2. Are they the same? Does this angle tell you anything about the actual distance between the barn and house or Moon and star?

We are unable to **directly** measure distances to objects in our night sky. However, we can obtain the distances to relatively nearby stars by using their parallax angles. Because even these stars are very far away (up to about 500 parsecs), the parallax angles for these stars are very small. They are measured in units of **arcseconds**, where 1 arcsecond is 1/3600 of a degree. To give you a sense of how small this angle is, the thin edge of a credit card, when viewed from 1 football field away, covers an angle of about 1 arcsecond.

Part II: Finding Stellar Distance Using Parallax

Consider the star field drawing shown in Figure 1. This represents a tiny patch of our night sky. In fact, the angle separating Stars A and B is just 1/2 an arcsecond.

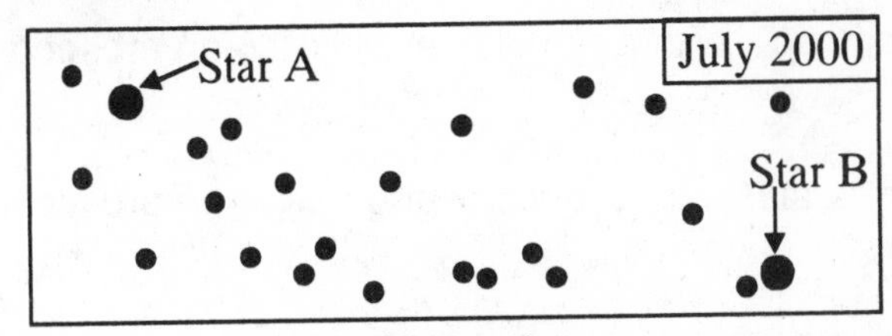

Figure 1

In Figure 2 (see the final page of the activity) there are pictures of this star field taken at different times during the year. One star in the field exhibits parallax as it moves back and forth across the star field with respect to the other more distant stars.

4) Using Figure 2, determine which star exhibits parallax. Circle that star on each picture in Figure 2.

5) In Figure 1, draw a line that shows the range of motion for the star exhibiting parallax in the pictures from Figure 2. Label the endpoints of this line with the months when the star is at those endpoints. What is the distance between the two end points in centimeters?

6) What is the distance between Stars A and B in Figure 1? Measure that distance in centimeters as well.

7) Recall that Stars A and B have an angular separation of ½ an arcsecond in Figure 1. Consider two more stars (C and D) that are separated by **twice** as many centimeters as Stars A and B. What is the angular separation between Stars C and D in arcseconds?

8) What is the angular separation between the endpoints that you marked in Figure 1 for the nearby star exhibiting parallax?

Recall that a star's parallax angle is defined as ***half*** the angular separation between the endpoints of the star's angular motion.

9) What is the parallax angle for the nearby star from question 8?

Recall that **one parsec** is defined as the distance to an object that has a **par**allax angle of one arc**sec**ond. A star with a parallax angle of 2 arcseconds would be ½ a parsec away.

10) For a star with a parallax angle of ½ an arcsecond, what is its distance from us?

11) For a star with a parallax angle of ¼ an arcsecond, what is its distance from us?

12) What is the distance from us to the nearby star exhibiting parallax in the pictures from Figure 2? (Hint: consider your answer to question 9.)
   a) 1 parsec
   b) 2 parsecs
   c) 4 parsecs
   d) 8 parsecs
   e) 16 parsecs

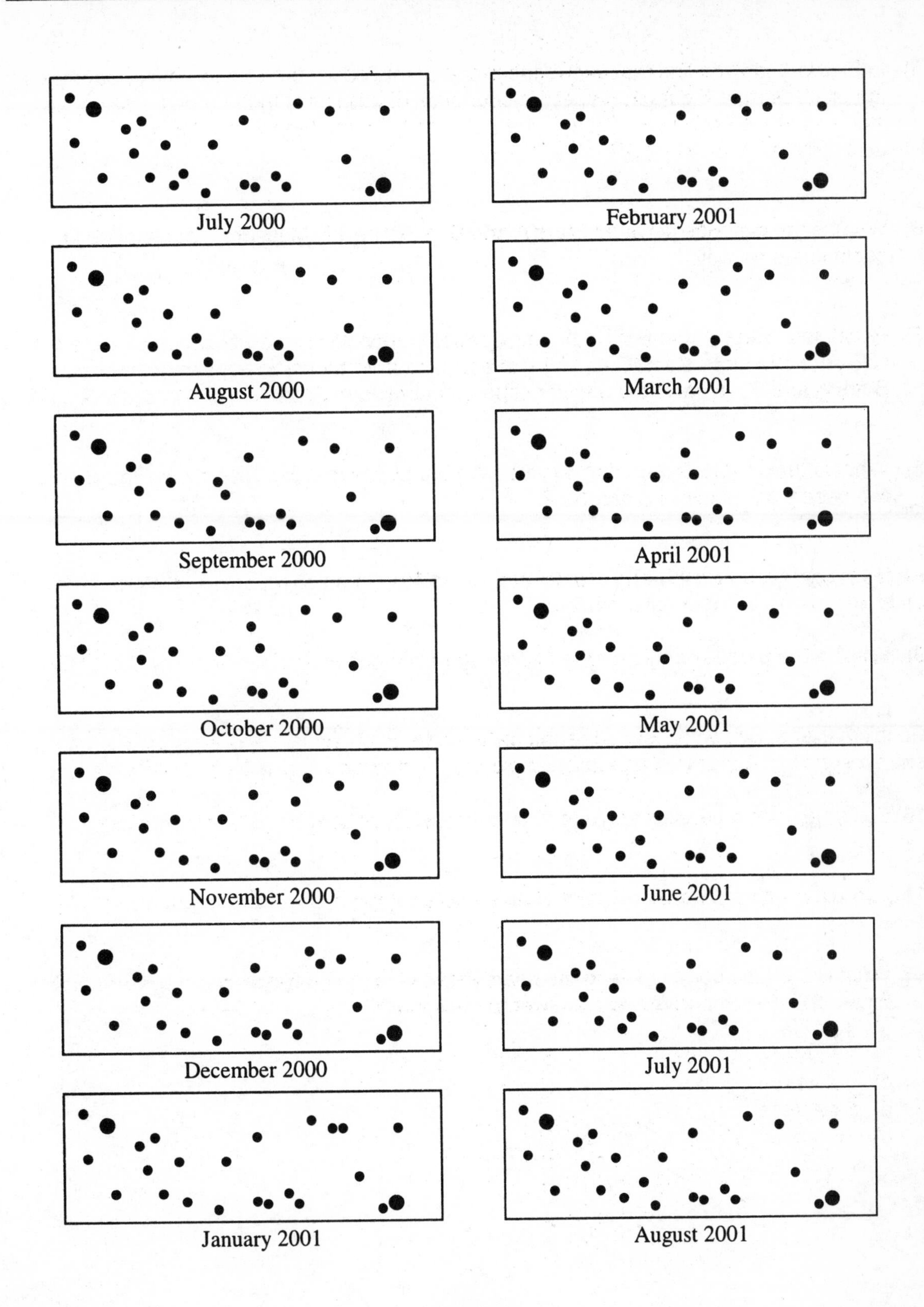
July 2000
February 2001
August 2000
March 2001
September 2000
April 2001
October 2000
May 2001
November 2000
June 2001
December 2000
July 2001
January 2001
August 2001

Figure 2

The H-R diagram below will be used to answer questions throughout this activity.

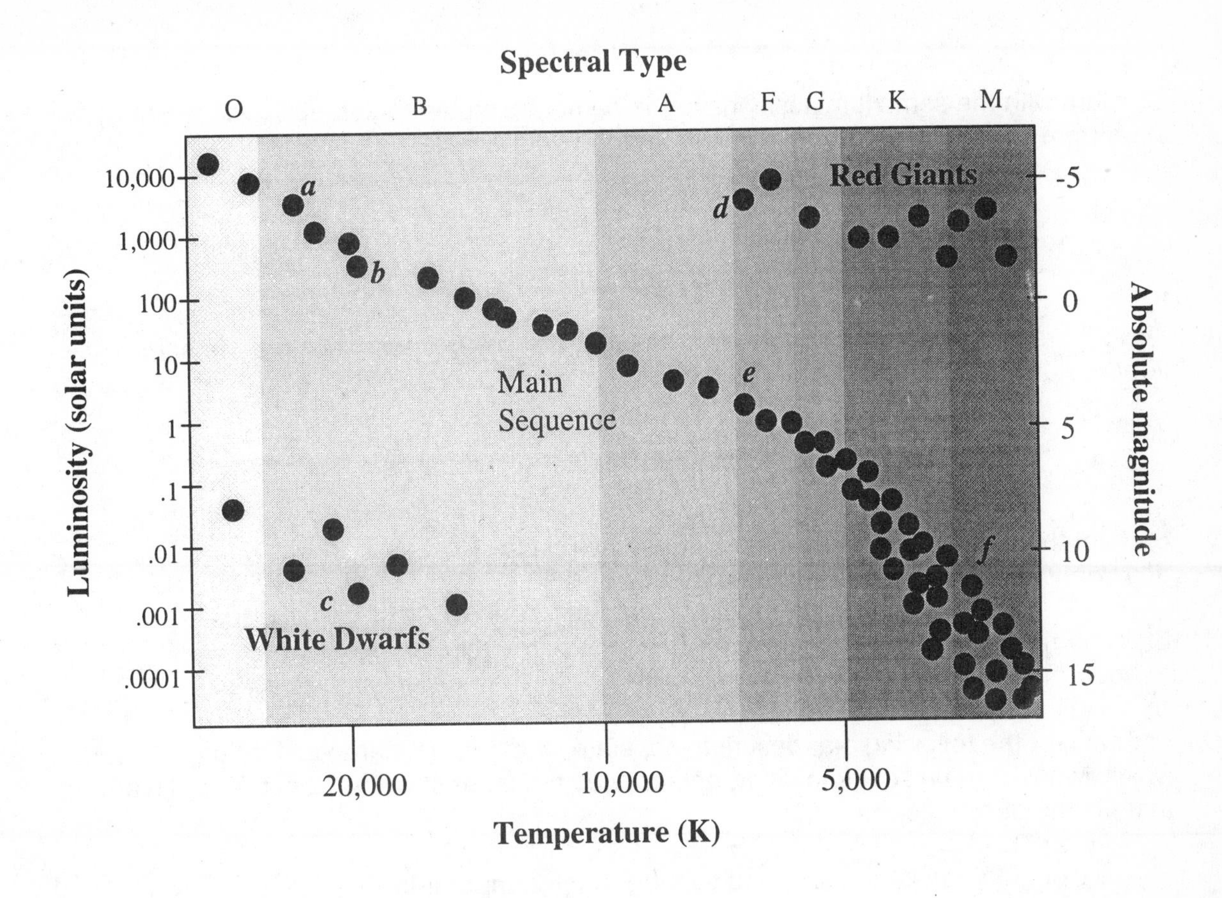

1) What are the spectral type, temperature, absolute magnitude, and luminosity of Star *a*?

Spectral type:

Temperature:

Absolute magnitude:

Luminosity:

2) Which 2 pairs of labeled stars in the diagram have the same temperature?

3) Do stars of the same temperature have the same spectral type? Use a pair of stars from your answer to question 2 to support your answer.

4) Which 2 pairs of labeled stars have the same luminosity?

5) Do stars with the same luminosity have the same absolute magnitude? Use a pair of stars from your answer to question 4 to support your answer.

6) If two stars have the same absolute magnitude, do they necessarily have the same temperature?

7) Stars of the same spectral type have the same (circle):
absolute magnitude *temperature* *luminosity*

8) Stars of the same absolute magnitude have the same (circle):
temperature *luminosity* *spectral type*

9) For each of the following star descriptions, state whether the star would be a red giant, white dwarf, or main sequence star, and provide the letter of a star from the H-R diagram that fits the description.

a) very bright (high luminosity) and very hot (high temperature)

b) very dim and cool

c) very dim and very hot

d) very bright and cool

Part I: Magnitudes and Star Distances

Below is a table of four stars and their apparent and absolute magnitudes. Use this table to answer the following questions.

	Apparent Magnitude	*Absolute Magnitude*	*Distance*
Star A:	0	0	
Star B:	0	2	
Star C:	5	4	
Star D:	4	4	

1) Which object appears brighter from Earth: Star C, Star D, or neither? Explain your reasoning.

2) Which object is more luminous: Star C, Star D, or neither? Explain your reasoning.

3) How far away is Star D? Record your answer in the table.

4) Use your answers from questions 1-3 to determine whether Star C is *more than 10 pc*, *less than 10 pc*, or *exactly 10 pc* away. Record your answer in the table.

5) Star B has an apparent magnitude of 0, which tells us how bright it appears from Earth at its true location. Star B has an absolute magnitude of +2, which tells us how bright it would appear if were at a distance of 10 parsecs (about 33 light-years).

   Where would Star B appear brighter, in its true location or if were at a distance of 10 parsecs? Explain your reasoning.

6) So, is Star B's location closer or farther than 10 parsecs? Record your answer in the table and explain your reasoning.

7) Complete the remaining blank in the above table and check your answers with another group.

Part II: Spectroscopic Parallax

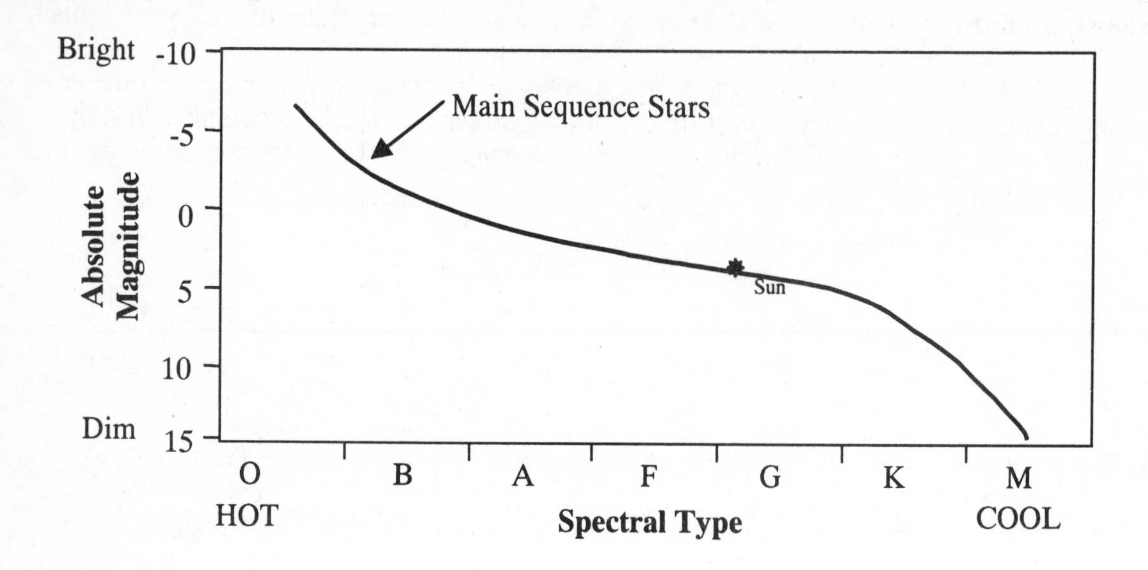

Below is a table giving both the apparent magnitude and spectral type for five **main sequence** stars. For each star, do the following:

8) Using the above H-R diagram, estimate the absolute magnitude for each star.

9) Complete the table below by classifying each star as being **closer**, **slightly farther**, or **much farther** than 10 parsecs away. This procedure, called spectroscopic parallax, provides astronomers with another strategy to measure the distance to stars.

Star	Apparent Magnitude	Spectral type	Absolute Magnitude	Distance Estimate
Rigel Kentaurus	0.0	G2		
Vega	0.04	A0		
Rigel B	6.6	B9		
Achernar	0.5	B3		
Tau Scorpius	2.8	B0		

Note: *By completing this table, you have estimated the distance to a star by comparing the apparent and absolute magnitudes without using a formula. The exact distance can be calculated using the formula* $d=10^{(m-M+5)/5}$ *pc, where m is the apparent magnitude and M is the absolute magnitude.*

Stars (like the Sun) begin life as a cloud of gas and dust. The birth of a star begins when a disturbance, such as the shockwave from a supernova, triggers the cloud of gas and dust to collapse inward.

1) Imagine that you are observing the region of space where a cloud of gas and dust is beginning to collapse inward to form a star (this region is called a protostar). Will the atoms in the collapsing cloud move away from one another, move closer to one another, or stay at the same locations?

2) What physical interaction or force causes the behavior described in question 1 to occur?

3) Would you expect the temperature of the protostar to increase or decrease with time? Explain your reasoning.

The continuous inward collapse of material causes the center of the protostar to become very hot and dense. Once these temperatures and densities reach a critical level, nuclear fusion reaction begins. During this fusion reaction, hydrogen atoms are combined together to form a helium atom. When this happens, a photon of light is emitted. Once these reactions are sustained, the protostar becomes a main sequence star (like the Sun). At this time the star no longer collapses, and a state of hydrostatic equilibrium is reached between the inward gravitational collapse of material and the outward pressure caused by the photons of light emitted as part of the nuclear fusion reaction.

Consider the information shown in the table below when answering questions 4 through 6.

Mass of the Star (*in multiples of Sun masses,* M_{Sun})	Main Sequence Lifetime of the Star
.67 M_{Sun}	45 billion years
1 M_{Sun}	10 billion years
1.3 M_{Sun}	800 million years
2 M_{Sun}	500 million years
6 M_{Sun}	70 million years
60 M_{Sun}	800 thousand years

4) Which live longer: high mass or low mass stars?

5) Based on your answer to question 4, do you think that the nuclear fusion rate for a larger mass star is greater than, less than or equal to the nuclear fusion rate of a low mass star?

6) The nuclear fusion rate of a star depends on both the temperature and the density of the star's core. How would the temperature and density of the core for a large mass star compare to that of a low mass star?

7) Which of the following statements best describes how the lifetimes compare between a star that has a mass equal to the Sun and a star with six times the mass of the Sun? Circle one of the responses given below.

A star with a mass the same as the Sun will:

a) live less than six times as long as a star with a mass six times the mass of the Sun.
b) live six times shorter than a star with a mass six times the mass of the Sun.
c) have the same lifetime as a star with a mass six times the mass of the Sun.
d) live six times longer than a star with a mass six times the mass of the Sun.
e) live more than six times longer than a star with a mass six times the mass of the Sun.

Explain your reasoning for the choice you made.

8) The Sun has a lifetime of approximately 10 billion years. If you could determine the rate of nuclear fusion for a star with twice the mass of the Sun, which of the following would it be? Circle one of the responses given below.

a) less than the nuclear fusion rate of the Sun
b) a little more than the nuclear fusion rate of the Sun
c) twice the nuclear fusion rate of the Sun
d) more than twice the nuclear fusion rate of the Sun

Explain your reasoning for the choice you made.

Main sequence stars that have used up their available hydrogen fuel supply become red giant stars. Although most main sequence stars become red giants, their specific evolutionary paths after this red giant phase vary greatly depending on mass.

A low-mass star, less than about 8 times the mass of our Sun, eventually ejects its outer layers to produce a planetary nebula. The stellar core remaining in the middle of this planetary nebula is called a white dwarf star. If the white dwarf is isolated, it will slowly cool until it becomes a black dwarf.

By contrast, a high-mass star will eventually explode as a Type II supernova. Depending on the original mass of the star, the Type II supernova will leave behind either a neutron star or, if the original star was extremely massive, a black hole.

1) Use the information above and the word list below to fill in the ovals in the diagram on the next page. Be sure to look at the arrows and words between the ovals to make sure these links between ovals make sense. Check your work with another group.

   <u>Word list:</u>
   neutron star
   black hole
   planetary nebula
   white dwarf
   nova
   Type II supernova

The diagram does not give us all the information known about the death of stars. Since it is incomplete, we can always add to this diagram when we learn more information.

2) In parts a and b below, you are given some additional information about the end states of stars. It is your job to change or add to the diagram to incorporate this additional information. (Note: There are several ways to accomplish this.)

   a) If a white dwarf has a nearby companion star, it can gravitationally attract material from its companion, a process known as accretion. When the white dwarf accretes enough material from the companion, the white dwarf will either explosively burn-off the outer layers as a nova, leaving behind the white dwarf unchanged, or explode as a Type I supernova, leaving nothing behind.

   b) In isolated space, a black hole can be nearly impossible to detect. However, if a black hole has a binary companion star, the strong gravitational pull of the black hole can accrete matter from its companion. This material spirals around the black hole. This process causes the rapidly moving material to emit large amounts of X-ray radiation, which we can detect with X-ray telescopes. Thus, one way to look for black holes is to look for strong X-ray sources.

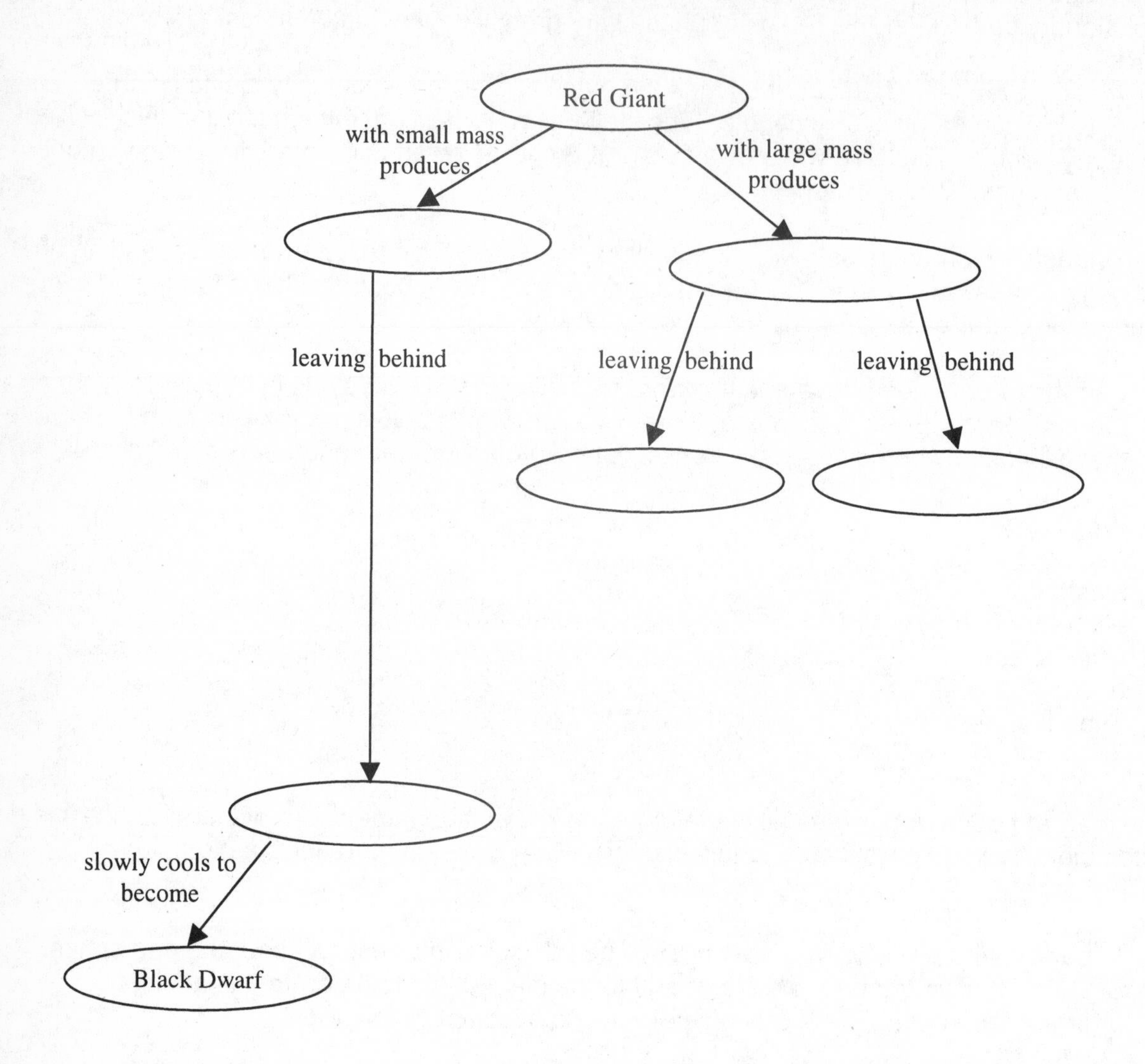
Red Giant
with small mass produces
with large mass produces
leaving behind
leaving behind
leaving behind
slowly cools to become
Black Dwarf

This tutorial will give you a better understanding of the size of the Milky Way Galaxy as it relates to other distances in the Universe. Below is a picture of a spiral galaxy. This is a picture of NGC 3184, a galaxy similar to the Milky Way. Because we are located within the Milky Way, we are unable to take a picture of our entire galaxy. Let's assume that this picture represents our Milky Way Galaxy, with the dimensions labeled below. **Note that in this picture, 1 centimeter represents 10,000 light years (ly).**

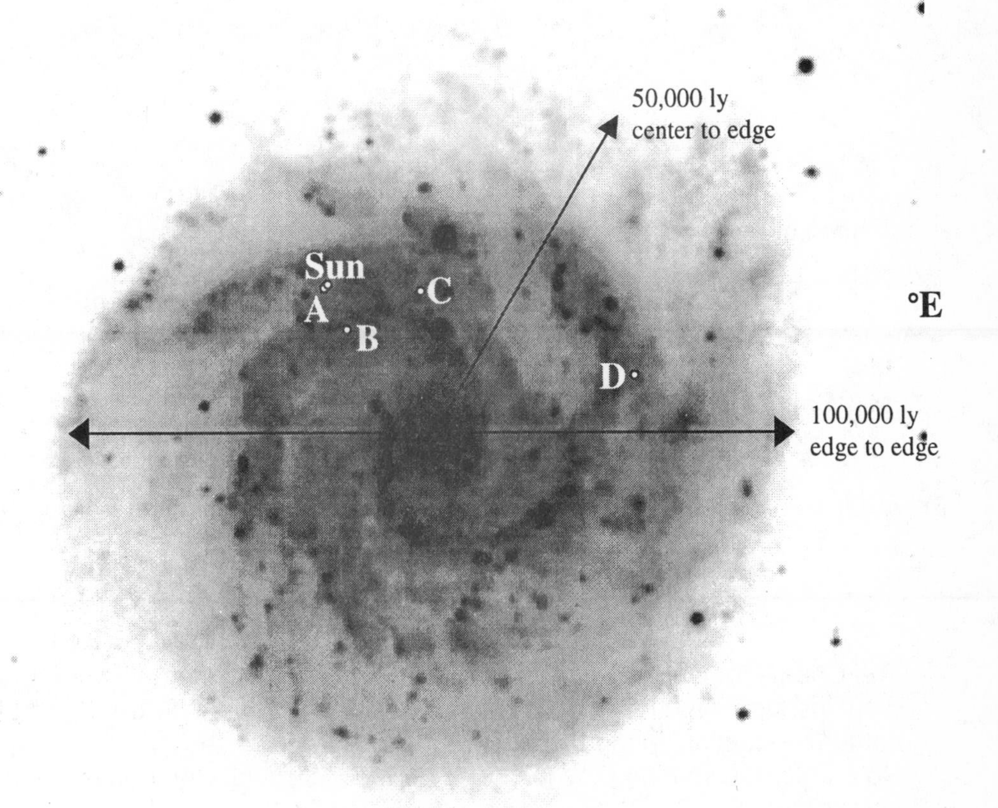

1) The Sun's position in the Milky Way is shown in the picture above. What is the distance from the Sun to the center of the Milky Way? Note: use a ruler and the fact that 1 cm = 10,000 ly.

2) The table below lists five bright stars in the night sky. Write the letter of the dot (A through E) on the picture above that best represents the location of each star. You can use letters more than once.

Star	Distance from Sun (in light years)	Letter
Sirius	9	
Vega	26	
Spica	260	
Rigel	810	
Deneb	1,400	

3) We normally consider Deneb to be a bright but distant star at 1,400 ly away. Compared to the size of our galaxy, is Deneb truly distant? Explain.

4) Are the stars from question 2 inside or outside the Milky Way Galaxy? Explain your reasoning.

5) The table below lists three Messier Objects and their distances from the Sun. Write the letter of the dot (A through E) on the picture above that best represents the location of each object. You can use letters more than once.

Messier Object	Distance from Sun (in light years)	Letter
M45 Open Cluster (Pleiades)	380	
M1 (Crab Nebula)	6300	
M71 Globular Cluster	12,700	

6) Are these Messier Objects part of the Milky Way? Explain your reasoning.

7) The Crab Nebula has a width of about 11 light years. If you wanted to accurately draw the Crab Nebula on your diagram, would you use a crab-shaped blob or a tiny dot at the location you indicated? Explain your reasoning.
Note: the dots marking the locations on the picture are about a tenth of a centimeter across.

8) The Sun is much smaller than a nebula. We used a dot to represent the Sun's location in the picture. Is this dot too small, too large, or just the right size to represent the size of the Sun on the picture? Explain your reasoning.

9) The Milky Way Galaxy is one of the largest galaxies in a group of nearby galaxies called the Local Group. The following table lists the distances to the centers of three Local Group galaxies. Draw a dot on your picture (if possible) to represent the center of each galaxy. Don't worry about the direction (left/right/up/down) for each galaxy; just place a dot an appropriate distance from the Sun.

Galaxy	**Distance from Sun (in light years)**
Sagittarius Dwarf Elliptical Galaxy (SagDEG) - closest galaxy to Milky Way	80,000
Large Magellanic Cloud	160,000
Andromeda Galaxy (M31)	2,500,000

Do any of these galaxies fit on the page? Which ones?

10) The objects in question 9 are all visible in the night sky from Earth. Are these objects inside or outside the Milky Way? Explain your reasoning.

11) SagDEG is approximately 11,000 ly across. Is this galaxy better represented on your diagram by a blob or a tiny dot? Explain your answer, and make an appropriate sketch to represent the galaxy.

12) Within the Local Group, the two largest galaxies are the Milky Way and Andromeda galaxies. From question 9, we saw that the Andromeda Galaxy was about 2,500,000 ly from us. On the picture, this spot would be 250 cm (about two and a half meter sticks) away from the dot representing the Sun.

The nearest group of galaxies to us (not counting our own Local Group) is the Virgo Cluster, about 60,000,000 ly away. How many centimeters away would this cluster be on our picture?

Imagine that you have received six pictures of six different children who live near six of the closest stars to the Sun. Each picture shows a child on his or her 12th birthday. The pictures were each broadcast directly to you (using a satellite) on the day of the child's birthday. Note the abbreviation "ly" is used below to represent a light-year.

John lives on a planet orbiting Ross 154, which is 9.5 ly from our Sun.
Peter lives on a planet orbiting Barnard's Star, which is 6.0 ly from our Sun.
Celeste lives on a planet orbiting Sirius, which is 8.6 ly from our Sun.
Savannah lives on a planet orbiting Alpha Centauri, which is 4.3 ly from our Sun.
Inga lives on a planet orbiting Epsilon Eridani, which is 10.8 ly from our Sun.
Ron lives on a planet orbiting Procyon, which is 11.4 ly from our Sun.

1) Describe in detail what a light-year is. Is it an interval of time, a measure of length, an indication of speed, or some other quantity?

2) Which child lives closest to the Sun? How far away does he or she live?

3) What was the greatest amount of time that it took for any one of the pictures to travel from the child to you?

4) If each child was 12 years old when he or she sent his or her picture to you, how old was each of the children when you received their picture?

John_____ Peter_____ Celeste___

Savannah _____ Inga _____ Ron _____

5) Is there a relationship between the current age of each child and his or her distance away from Earth? If so, describe this relationship.

6) Imagine that the six pictures were broadcast by satellite to you and that they arrived at exactly the same time. For this to be true, does that mean the all of the children sent their pictures at the same time? If not, which child sent his or her picture first and which child sent his or her picture last?

7) The telescope image at the right was taken of the Andromeda Galaxy, which is located 2.5 million ly away from us. Is this an image showing how the Andromeda Galaxy looks right now, how it looked in the past, or how it will look in the future? Explain your reasoning.

8) Imagine that you were observing a distant star that was located in a galaxy 100 million ly away from you. By analysis of the starlight received, you are able to tell that the image we see is of a 10 million year old star. You are also able to predict that the star will have a lifetime of 50 million years, at which point it will end in a catastrophic supernova.

   a) How old does the star appear to us here on Earth?

   b) How long will it be before we receive the light from the supernova event?

   c) Has the supernova already occurred? If so, when did it occur?

9) Imagine that you take two images of two main sequence stars that have the same mass. From your observations, both stars appear to be at the same point in their evolution. Consider the following possible interpretations that could be made from your observations.

   a) Both stars are the same age and the same distance from you.
   b) Both stars are the same age but at different distances from you.
   c) The stars are actually different ages but at the same distance from you.
   d) The star that is closer to you is actually the older of the two stars.
   e) The star that is farther from you is actually the older of the two stars.

   How many of the five choices (a-e) are possible? Which ones? Why?

The two drawings below represent the same group of galaxies at two different points in time during the history of the Universe.

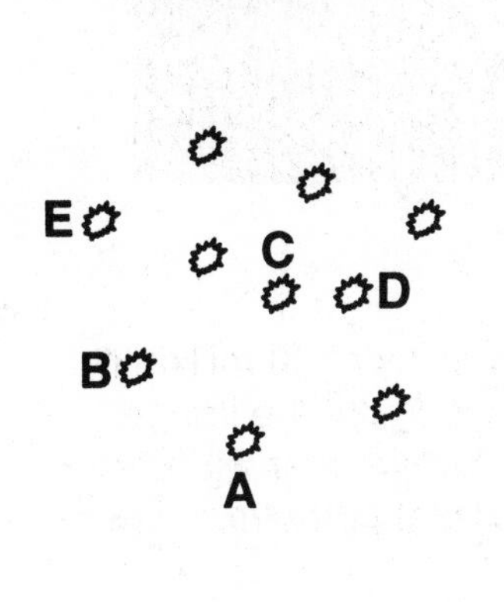

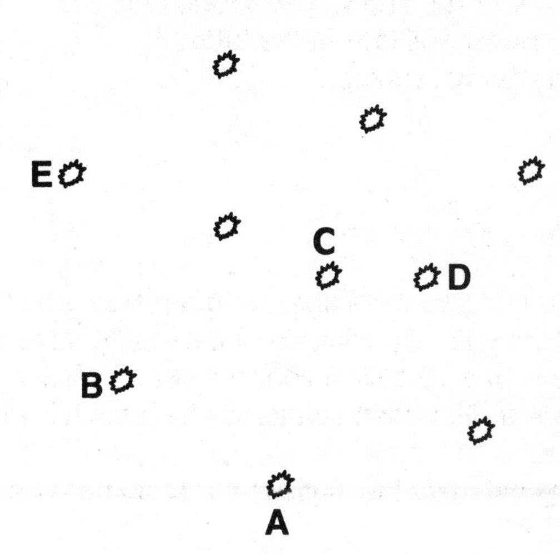

1) Examine the distance between the galaxies labeled A - E in the *Early Universe*. Are all the galaxies the same distance from each other?

2) Describe how the Universe changed in going from the *Early Universe* to the *Universe Some Time Later*.

3) Do the galaxies appear to get bigger?

4) Will stars within a galaxy move away from one another due to the expansion of the Universe?

5) Compare the amount that the distance between the D and C galaxies changed in comparison to the amount that the distance between the D and E galaxies changed. Which galaxy appears to have moved farther, C or E?

6) If you were in the D galaxy, how would the A, B, C and E galaxies appear to move relative to your location?

7) If you were in the D galaxy would the A, B, C, and E galaxies all appear to move by the same amount in the time interval from the *Early Universe* to the *Universe Some Time Later*?

8) Rank the A, B, C, and E galaxies (from greatest to smallest), in terms of their relative speeds away from you in the D galaxy.

9) Now imagine that you are in the E galaxy. Rank the A, B, C and D galaxies, in terms of their relative speeds away from you, from greatest to smallest.

10) Is there a relationship between the apparent speed of an object and its distance from you in the Universe? If so, describe this relationship.

11) Would your answer to question 10 be true in general for all locations in the Universe?

12) Consider the following discussion between two students regarding the possible location of the center of the Universe.

Student 1: *Since all the galaxies we observe are moving away from us we must be at the center of the Universe.*

Student 2: *If our Milky Way Galaxy were like galaxy A and if the Andromeda galaxy were like galaxy B, then we would both see all other galaxies moving away from us. So I'm not sure if our Milky Way Galaxy is the center or if it's Andromeda.*

State whether you agree or disagree with *each* of the student statements. Explain your reasoning for *each*.